Post-harvest Technology of Fish and Fish Products

Post-harvest Technology of Fish and Fish Products

K.K. Balachandran

2025

Daya Publishing House®

A Division of

Astral International Pvt. Ltd.

New Delhi - 110 002

First Published, 2001
Reprinted, 2025

ISBN: 978-81-7035-237-2 (Hardbound)

Publisher's note:

Every possible effort has been made to ensure that the information contained in this book is accurate at the time of going to press, and the publisher and author cannot accept responsibility for any errors or omissions, however caused. No responsibility for loss or damage occasioned to any person acting, or refraining from action, as a result of the material in this publication can be accepted by the editor, the publisher or the author. The Publisher is not associated with any product or vendor mentioned in the book. The contents of this work are intended to further general scientific research, understanding and discussion only. Readers should consult with a specialist where appropriate.

Every effort has been made to trace the owners of copyright material used in this book, if any. The author and the publisher will be grateful for any omission brought to their notice for acknowledgement in the future editions of the book.

Published by : **Daya Publishing House®**
A Division of
Astral International Pvt. Ltd.
– ISO 9001:2015 Certified Company –
4723/23, Ansari Road, Darya Ganj
New Delhi-110 002
Ph. 011-43549197, 23278134
E-mail: info@astralint.com
Website: www.astralint.com

Laser Typesetting : **Classic Computer Services**

Printed at : **Replika Press Pvt. Ltd.**

FOREWORD

LAST few decades had witnessed widespread development in the technologies for capture as well as culture of aquatic organisms world over. There is tremendous advancement in post-harvest technology in the field of fisheries. Improvements in pre-process handling, processing, packaging and transportation are quite appreciable, even in developing countries like India. Being a commodity of high economic value earning substantial quantum of foreign exchange, the fish and fishery products receive utmost care and importance and the scope for further development in this sector is quite promising. The policy of importing raw material from other countries for processing, which in turn helps in utilisation of the under-utilised built-up capacity of the processing industry, will open up new possibilities.

Human resources development in this sector is being well attended by a chain of fisheries colleges, functioning under the State Agricultural Universities, Fisheries Institutes like Central Institute of Fisheries Education, Central Institute of Fisheries Technology, Central Institute of Freshwater Aquaculture and Central Marine Fisheries Research Institute functioning under the Indian Council of Agricultural Research and, Central Institute of Fisheries Nautical, Engineering and Training, under the Government of India.

Lack of quality textbooks is widely felt, especially based on Indian conditions and context. The book entitled *Post-harvest Technology of Fish and Fish Products* by Dr. K.K. Balachandran, Principal Scientist, Central Institute of Fisheries Technology is one which I am sure will be able to solve this problem to a great extent. Dr. Balachandran with his long experience in this field, and very wide knowledge in all aspects of post-harvest technology, has dealt with these topics in much detail in different chapters. This book will be a good textbook for under-graduate and post-graduate studente of fisheries and technologists in the field of fish processing, and a reference book for scientists as well as those interested in fish processing industry.

Dr. D.M. Thamphy
Dean, College of Fisheries
Kerala Agricultural University

PREFACE

FISHERIES is a vital sector contributing substantially to the Indian economy. It is a major provider of employment, next only to agriculture, and the much-needed inexpensive wholesome protein food to the masses. It is, therefore, no wonder that 'fisheries' has been given due importance in the developmental activities of India.

'Fisheries' is a multi-disciplinary field. Even with the best of infrastructure in terms of facilities and technology, the success of operations will depend on the qualified manpower behind them. This fact has been very well understood and the necessary infrastructure has been developed by way of establishing several teaching and training institutions in fisheries all over the country. Commencement of educational programmes under State Agricultural Universities with facilities for graduate, post-graduate and doctoral programmes in different disciplines of fisheries is the most welcome move in this direction. Post-graduate and doctoral programmes are being offered also from different central fisheries research organisations under the Indian Council of Agricultural Research (ICAR). There is a full fledged institution for fisheries education alone with the status of a Deemed University functioning under the ICAR. Courses in different disciplines of fisheries at graduate levels are also offered at several colleges in the country affiliated to other regular universities. Fisheries courses have been started even at the higher secondary levels in some States.

Though the institutions offering post-graduate and doctoral courses in fisheries may have the adequate support of library facilities, such facilities are often insufficient or even totally absent in other places. This is a major hurdle faced by the students and the teaching faculty alike, more so at the higher secondary and graduate levels. The probable remedy in such a situation is the availability of textbooks comprehensively covering different disciplines for different courses. It is with a view to fill up this gap in this much needed area in fisheries education that an attempt has been made to prepare a book on fish processing technology.

Though the book is intended to serve as a textbook, it is not prepared on the basis of the syllabus of any particular educational course. Therefore different topics are treated in such a way as to provide a comprehensive knowledge of the theoretical as well as the applied aspects involved. The author will be happy to receive critical comments and suggestions from the students, teachers and other users of the book to improve its quality further.

K.K. Balachandran

ACKNOWLEDGEMENTS

I am indebted to several of my colleagues at the Central Institute of Fisheries Technology and friends in the teaching profession for earnestly entreating me to write this book, going through the manuscript and, offering critical comments to improve the contents. Of particular mention among them are Drs. P. Madhavan, T.S. Unnikrishnan Nair, A.C. Joseph, P.K. Vijayan, K. Devadasan, P.K. Surendran, P. Ravindran and M.M. Devasya. I am also deeply indebted to K.P. Leelamma for her excellent secretarial assistance and the work on the computer to set the manuscript in the printable form. The assistance in graphics by B. Sudhir also is gratefully acknowledged. I am also thankful to Dr. D.M. Thampy, Dean, College of Fisheries, Kerala Agricultural University, for writing a Foreword to the book. I also wish to place on record my thanks to M/s. Daya Publishing House for the excellent printing and publication within a short time.

K.K. Balachandran

CONTENTS

1
BIOCHEMISTRY AND NUTRITION

PRESERVATION and processing invariably bring about some changes, most of them irreversible, in the intrinsic characteristics of the fish. It is a fact that consumers always prefer fresh to preserved fish. The lack of preference for preserved or processed fish is attributed, to some extent, to the changes in the appearance, colour, texture, taste and flavour accompanying the processing involved. The severity as well as the irreversibility of the changes will depend on the type of the preservation/processing technique the fish has been subjected to. The developments in fish processing technology over the years have been oriented towards presentation of preserved/processed products with characteristics closely resembling those of the fresh fish. From the traditional methods of drying, salting etc. the technology of preservation has advanced to canning, freezing and several others, transportation of live fish being the latest.

Fish being one of the most perishable among foodstuffs, processing aims at controlling, if not totally arresting, the process of spoilage and make the fish available in a variety of forms acceptable to the consumer. The biochemical changes taking place in the fish *post-mortem* is very complex. Several changes take place in the fish muscle constituents leading to changes in texture and flavour producing odoriferous compounds indicative of spoilage. The degree of spoilage is dependent on several factors, some of which are intrinsic, that is the sum of attributes inherent in the fish muscle. Besides, there are several external factors having direct bearing on spoilage. A thorough understanding of these factors and their roles in the mechanism of spoilage of fish is essential to maintain spoilage under control and to develop a processed product with maximum acceptability to the consumer.

1.1 PROXIMATE COMPOSITION OF FISH FLESH

By proximate composition is meant the composition of the major constituents. In fish the major constituents are water, protein, lipids and ash. In addition to these major constituents, fish also contains some minor constituents like non-protein nitrogen compounds, vitamins and carbohydrates. Large variations occur in the proximate composition and are influenced by several factors like the species of the fish, diet, fishing grounds, season, sex and sexual maturity, and spawning. Considerable variations in the proximate composition have been found to occur between fish of the same species also.

1.1.1 Water

Water is the principal constituent of fish, quantity-wise. Water content in many fish and shellfish varies between 60 and 80 per cent. Water content as high as 90 per cent is met with in Bombay duck (*Harpodon nehereus*). Water content is also found to vary considerably within the same species of fish depending on the age, fat content, feeding condition, spawning etc. Fatty fish exhibit an inverse relationship between fat and water contents.

1.1.2 Protein

The characteristic properties of fish, which determine the degree of its excellence, include both wholesomeness and sensory characteristics. An important criterion determining the wholesomeness is the nutritive value of proteins. Protein is the most important constituent of fish from the nutritional point of view. Protein content in most fish averages 18-20 per cent, though the general variation is in the range 15-24 per cent. Compared to finfish species, shellfish fall in a lower category as far as protein content is concerned, the range being 8-15 per cent. In some shellfish like oyster, values in the range 5-14 per cent are common. Variations in protein content occur in relation to age, fat content, spawning, starvation etc.

Proteins are highly complex nitrogenous organic substances of very high molecular weight. They are colourless, amorphous and colloidal in nature. Proteins are polymers of amino acids that contain the elements carbon, hydrogen, oxygen nitrogen and, in some cases, sulphur. The amino acids are united by a peptide linkage –CO-NH-. Protein is a necessary dietary intake for animals since they cannot synthesise them in their body. In the muscles and different organs of animals, proteins perform their functions in association, or interacting, with other components like water, lipids, minerals etc.

1.1.2.1 Protein quality

Proteins are made up of about 20 amino acids. Proteins in the diet serve as the source of these amino acids for the synthesis of human body protein. Eight to ten of these amino acids cannot be synthesised in the human body, or they cannot be synthesised in adequate quantity to meet the needs of optimal rates of growth in children. They have to be provided through food and are called essential amino acids. They must be present in adequate quantities in the protein to make it of high nutritional quality. If one or more of these amino acids are absent or are insufficient in the intake food, the protein synthesis in the body will be seriously impaired leading to non-utilisation of protein. This will result in reduced growth in children or loss of muscle mass in adults. Protein malnutrition is often described in its two extreme forms - marasmus, a generalised wasting due to a deficiency of both protein and energy and, kwashiorkar characterised by edema and general deficiency.

The essential amino acids required in the diet are:

- lysine
- methionine
- threonine
- tryptophan
- leucine
- isoleucine
- valine
- phenylalanine
- arginine
- histidie

However, a dietary supply of arginine is not required for the adults.

Vegetable proteins are generally deficient in essential amino acids compared to animal proteins, whereas fish proteins contain all the essential amino acids, most of them in quantities sufficient or more than the dietary requirements. An exception is that of methionine or cystine which are present at levels slightly below the dietary needs. A protein is considered only as useful as the essential amino acid in the smallest quantity. The essential amino acids in the proteins not sufficient to meet the dietary needs and occurring furthest below the standard requirements are called limiting amino acids.

The other amino acids commonly found in proteins are alanine, arginine, aspartic acid, cysteine, glutamic acid, glycine, proline, serine and tyrosine.

Fish protein is a very rich source of the essential amino acid lysine that is absent in vegetable proteins. Fish protein can, therefore, be used to supplement vegetable proteins to provide a nutritionally balanced protein source in the diet.

Fish proteins can be generally classified into three categories, sarcoplasmic (enzymic), myofibrillar (contractile, structural) and stroma (connective tissues) proteins. The grouping is so done because the constituents of different fractions are localised in different entities of the muscle cell and contribute in different fashions to its various activities. The difference in the extractability of the various fractions is one of the bases of such classification.

1.1.2.2 Sarcoplasmic proteins

Sarcoplasmic proteins or the 'myogen fraction' of the fish muscle is a large family of proteins and constitute 20-22 per cent of the total proteins of fish muscle. They include myoglobin, several enzymes and other albumins. They occupy the space between the myofibrils. They are characterised by their solubility in water and salt solutions of low ionic strength of the order of 0.05. They do not contribute significantly to the filamentous organisation of the muscles. Their functions are directed mainly to metabolic activities of the cell. The content of sarcoplasmic proteins is generally higher in pelagic fish and lower in demersal fish. In shellfish like shrimp, this fraction is generally very high, of the order of 30 per cent of the total proteins.

• ENZYMES

Most of the sarcoplasmic proteins are enzymes. The sarcoplasmic enzymes are involved in the physiological events of respiration, intracellular digestion, cell growth, cell division and secondary metabolism. The main enzyme groups identified are hydrolases, oxidoreductases and transferases.

The hydrolases include proteinases, peptidases, lipases, phospholipases, nucleotidases and glycogen hydrolase. Proteinases and peptidases are involved in the autolysis of the proteins and, lipases and phospholipases in the hydrolysis of lipids. Nucleotidases are responsible for the degradation of nucleic acids and glycogen hydrolase breakdown glycogen to lactic acid.

The oxidoreductase enzymes like phenoloxidase have been incriminated against development of black discolouration in shellfish like shrimp, lobster etc. and are of great significance in post-harvest handling of these shellfish species. Some enzymes of this class have been held responsible for the blackening of carotenoid pigments in the skin.

- HEME PROTEINS

Fish meat is generally white in colour except for the red muscle found located just under the skin along the side of the body in some fish. Active swimmers like tuna have considerable amount of black meat around the backbone. Dark muscles have higher levels of lipids, glycogen and most vitamins. Dark to white muscle proportion is related to activity. Continuous swimmers like sardine, mackerel etc. have higher content of dark muscle. Demersal species move only periodically and hence have less of dark muscle.

The pigments imparting the red surface colour of the fish meat include oxymyoglobin and oxyhaemoglobin. Haemoglobin will be lost during handling and storage, but the myoglobin will be retained in the cellular structure and contribute to the colouration. The fish myoglobins are readily oxidised to metmyoglobins.

Haemocyanins are copper-containing proteins found in the blood of crustaceans like shrimp, crab etc. Heat-denatured haemocyanin or oxyhaemocyanin can form a green-blue coloured compound which is of significance in canned shellfish like crab and shrimp.

Other soluble constituents of the sarcoplasmic fraction include various nitrogen-containing compounds like amino acids and nucleotides, soluble carbohydrates etc. This will be discussed in a separate section.

1.1.2.3 Myofibrillar Proteins

These are proteins responsible for the filamentous organisation of the muscle and are called structural proteins. The myofibrillar proteins account for 65-75 per cent of the total fish muscle proteins. They are characterised by their solubility in salt solutions of high ionic strength like 5 per cent sodium chloride or 0.1 N sodium hydroxide. The resistance to extraction is partly the result of intimate association and interaction between the proteins within the myofibrils. The high viscosity of the extract indicates the fibrous nature of the proteins brought into solution. Myofibrillar proteins are directly involved in muscle contraction, rigor mortis and, protein denaturation and the associated toughness of texture. They are also important in coagulation and gel formation in the fish muscle during cooking. The myofibrils progressively lose their extractability in weak salt solutions during freezing and cold storage rendering the fish muscles fibrous and tough.

The myofibrillar proteins present in the largest amounts in fish muscle are myosin, actin, tropomyosin and troponins. These are mainly involved in muscle contraction. Some minor constituents classified as scaffold proteins and regulatory proteins are also present which, though present at levels less than 1% each, contribute to the functional properties of the meat.

• MYOSIN

Myosin is a major fraction of the myofibrillar proteins. It is readily soluble in solutions of ionic strength higher than 0.3. The striated appearance of the myofibrils is due to the thin muscle filaments running between the thicker filaments. Myosin is the protein of the thick filaments. It contains two identical polypeptide chains, each with a helical structure. In addition, the two chains are supercoiled, i.e., wound around each other.

Myosin has a globular head which has the ATPase activity. This provides the ability to myosin to interact with actin. ATP molecules are also bound to the myosin filaments. Thus both ATP and ATP-ase are present in the myosin filament. Therefore, myosin is an enzyme and a fibrous protein at the same time.

Myosin constitutes approximately 50-60 per cent of the myofibrillar contractile proteins of the fish muscle. It can be extracted from the muscle with a salt solution of ionic strength 0.6 with a slightly alkaline pH. 0.3 M potassium chloride and 0.15 M phosphate at pH 6.5 are the solutions commonly used for extraction. Extracted myosin can be purified by repeated cycles of precipitation followed by resolubilisation in salt solution of high concentration.

• ACTIN

Actin is the major protein of the thin filaments. It constitutes 15-30 per cent of the myofibrillar proteins of the muscle. Actin is bound to the structure of the muscle more firmly than the myosin. Actin is extracted from the muscle powder obtained after extraction with acetone. For extraction, the muscle powder is exposed with an aqueous solution of ATP for a long period.

Actin exists in two different forms, the F actin and the G actin. Actin exists in the muscle as a double helical structure of the fibrous actin or F actin. Globular actin, G actin, is the monomeric form of the protein and is stable in water. In water it can also exist as a dimer. G actin binds ATP very firmly. In the presence of magnesium, G actin will spontaneously polymerise to form F actin with the concurrent hydrolysis of the bound ATP to yield bound ADP and inorganic phosphate. G actin also polymerises in the presence of neutral salts at concentration of approximately 0.15 M.

In the case of fish meat, large amounts of actin are also extracted with myosin. To extract actin from fish meat, minced meat is first made from the fish. It is then washed with 0.4 per cent solution of sodium bicarbonate and is then centrifuged. The residue is mixed with cool acetone to remove the water and make dried powder. This is the G actin. G actin is mixed with neutral salt and polymerised when it turns into F actin.

• ACTOMYOSIN

When actin is mixed with myosin *in vitro* a complex compound called actomyosin is formed. Though actin has no enzymic activity, it significantly modifies the ATPase activity of the myosin in the actomyosin complex. The complex of actomyosin can be dissociated in the presence of ATP and/or ADP and magnesium ions. It is quite probable that a similar type of combination between actin and myosin takes place in the muscle cells and is intimately involved in the muscle contraction. The interaction of actin and myosin in the absence of ATP and ADP, and the plasticising effect of these nucleotides when they are present, are important factors deciding the quality of the fish meat.

• TROPOMYOSIN

Tropomyosin is another important fibrous protein of the myofibril. It is a very well characterised fibrous protein. It is remarkably resistant to acid, alkali or heat treatment and is not easily denatured. It forms true crystals with sharp edges and faces. There is no enzymic activity associated with this protein.

Tropomyosin can be extracted from the fish muscle dehydrated with ethanol and ether and purified by precipitation from 1 M KCl at pH 7 and fractionation using 1.7-2.8 M ammonium sulphate solution.

Tropomyosin and troponin regulate muscle contraction. Biochemically, they are very important, though their content is very low. They do not play any role in food processing.

1.1.2.4 *Stroma Proteins*

Stroma proteins, which are insoluble in salt solution, constitute about 3 per cent of the proteins in teleosts and about 10 per cent in elasmobranchs. An essential difference between the fish meat and animal meat is that the fish flesh is tender while animal meat is often tough and hard to chew. The toughness of animal meat is due to its high content of connective tissues called stroma proteins. These tissues are composed of a highly hydrated amorphous ground substance in which several types of cells, as well as the fibrous proteins collagen and elastin are embedded. Collagen is the major component of the connective tissues. Fish contains very little connective tissues and thus very little collagen. Further, the collagen of fish becomes tender and soft on cooking.

In fish muscle the collagen fibrils are seen in the membranes enclosing each muscle fibre and muscle bundle and in the myocommata, i.e., the sheets separating the adjacent blocks of muscle fibres. The distribution of collagen in the muscle of fish is closely related to the swimming mode of the fish. The highest content is in the flexible musculature. The ordinary muscles of the dark-fleshed fish contain more collagen than the corresponding muscles of white-fleshed fishes. In starving fish, the sarcoplasmic and myofibrillar proteins undergo gradual depletion. However, the connective tissues remain unutilised.

Collagen is comparatively poor in essential amino acids and hence is not very important nutritionally. It is characterised by an unusual make up of amino acids in that nearly one third is made up by glycine. Collagen is unique also in that it is the only protein having a large content of the amino acid hydrxyproline and that it contains hydroxy lysine. It has also a large concentration of proline.

Collagen has a very profound influence on the texture and the rheological properties of the fish muscle and the products processed out of fish. There is direct correlation between the collagen content and the toughness of the fish muscle. Fish containing low amounts of collagen generally becomes tender, succulent and elastic after cooking due to the gelatinisation of the collagen.

1.1.3 Lipids

A factor showing wide variations in the fish is the lipid content. Lipids are a broad group of substances that have some properties in common and some similarities in composition. Fats and oils are the most important group of lipids. They are the simplest form of lipids and are esters of fatty acids, mainly fatty acid esters of glycerine and are called triglycerides. Compound form of lipids may contain other groups in addition to the esters of fatty acids with alcohol; e.g., phospholipids which are esters containing fatty acids, phosphoric acid and other groups usually containing nitrogen.

Basically, the lipid composition of the meat can be categorised as lipids from the muscle tissues and lipids from the adipose tissues. Lipid composition of the two tissues can be quite different. The lean portion contains greater proportion of phospholipids and these are located in the membrane of cell walls. Lean muscle contains about 0.5-1 per cent phospholipids. The fatty acids of phospholipids are more unsaturated that those of triglycerides. Consequently, fatty acids in the lean portion of the meat have a higher degree of unsaturation than those in the adipose tissues.

The term 'fat' is generally applied to all triglycerides, whether they are normally solid or liquid at ambient temperatures. Liquid fats are commonly referred to as oils.

Lipid content of the fish muscle shows wide variation from as low as 0.2 per cent to as high as 60-65 per cent in some species. Fat in the form of triglycerides is the most concentrated energy source of fish. Fatty components other than triglycerides like the phospholipids in the cell membrane, hormones, vitamins, carotenoids etc. compose of only a very small fraction of the total lipid content of fish.

Fish is often classified as lean, semi-fatty and fatty depending on the lipid content in the muscle. Fish with less than 0.5 per cent muscle lipids is categorised as lean fish; those with lipids in the range 0.5-2 per cent is semi-fatty and fish with more than 2 per cent lipids in the muscle is fatty fish. Unlike lean or semi-fatty fish, fat content in the fatty fishes exhibit wide fluctuations depending on growth, maturity, feeding, spawning etc. Fat is stored in fat tissues which are composed of special fat cells. Fat of true fish is found mainly in the dark muscle just below the skin and in the lateral line area. In some species there are extensive store of fat in the liver, in some others it is in the head and, in some in the peritoneum. Some fish have slightly higher fat content in the belly flap.

Fats, in addition to being the source of concentrated energy, also are carriers of fat-soluble vitamins. Fats also contribute to the taste, flavour and palatability of the fish, particularly the feeling of satiety after eating.

Fats are esters of glycerol with long chain fatty acids.

$$\begin{array}{lclcl}
CH_2\text{-}OH & & HOOC\text{-}R_1 & & CH_2\text{-}O\text{-}CO\text{-}R_1 \\
| & & | & & | \\
CH\text{-}OH & + & HOOC\text{-}R_2 & \rightarrow & CH\text{-}O\text{-}CO\text{-}R_2 + 3\,H_2O \\
| & & | & & | \\
CH_2\text{-}OH & & HOOC\text{-}R_3 & & CH_2\text{-}O\text{-}CO\text{-}R_3
\end{array}$$

R_1, R_2 and R_3 represent the hydrocarbon chain of the fatty acids which can be the same or different. In fish, many of the fatty acids present are unsaturated. Generally even numbered fatty acids occur in fish fatty acids, though odd numbered and branched chain fatty acids are not uncommon.

Fatty fishes exhibit great variations in the spatial distribution of fat in the body. A higher percentage of fat is always seen in the ventral side than in the dorsal. Similarly, the fat content can be seen to increase towards the tail region and also from the interior towards the exterior of the musculature. The higher fat in the tail region meets the increased needs of energy in swimming. Likewise the higher fat store in the ventral region favours to reduce the 'drag' during swimming. During starvation or spawning the fat is easily depleted from the musculature or liver as the case may be.

Fish which store fat in the musculature show an inverse relationship between fat and water contents. Fat and water together constitute approximately 80% of the fresh fish muscle. When the fat content increases the water content decreases and *vice versa*.

Fish fat is different from both vegetable and animal fats in their fatty acid make-up. Animal fats contain mainly saturated fatty acids. Vegetable fats may contain more unsaturated fatty acids than fish fat; however, the degree of unsaturation is greater in fish fat than in vegetable fat. In human nutrition some fatty acids like linoleic or linolenic acids are regarded as essential fatty acids (EFA) since they cannot be synthesised by the organism. EFA are essential food factors like essential amino acids or vitamins. In fish, EFA constitutes only around 2 per cent of the total lipids.

Fish fat is a source of highly unsaturated fatty acids having two or more double bonds called polyunsaturated fatty acids (PUFA). Among these, fatty acids which have their first double bond on the third carbon atom, i.e., the third from the carbon atom containing the methyl group, are called ω-3 or n-3 fatty acids. ω-3 PUFA has great medicinal value as they have anticholesterolemic and antithrombotic effects. The prominent among them are the eicosapentaenoic acid with five double bonds and docosahexaenoic acid with six double bonds. Fish oils like sardine oil is a good source of both these acids. Some PUFA in fish oils are known to cure skin diseases. ω-3 PUFA also have growth promoting effects in children.

A small portion of the lipids serves as essential structural parts of the cell. Typical 'non-depot' lipids are phospholipids which are phosphorus and nitrogen containing fats. They do not function as energy sources.

Nutritionally, fats contribute to the calorific value of fish. They provide 9 calories of energy per gram which is approximately double the energy provided by protein or carbohydrates. Fish with higher fat content will have higher calorific value. However, fat contents vary widely and no such fluctuations take place in the content of proteins. Therefore lean fish and shellfish will have highly consistent nutritional value compared to fatty fish.

Most of the tropical fish except the pelagic species like sardine, mackerel, tuna etc. belong to the low fat-high protein category.

1.1.4 Ash

Ash is the portion left after complete combustion of the organism. Ash content varies between 0.4 and 2% in fish. Ash is constituted by the minerals present in the fish. Minerals are inorganic substances required by the living organisms in many of their biological functions, e.g.

- calcium, phosphorus and magnesium are constituents of bone and teeth
- soluble salts of minerals like sodium, potassium, magnesium and phosphorus help to control the composition of body fluids and cells
- minerals like iron, phosphorus and most of the other elements are essential additives to many enzymes and other proteins necessary for the release and utilisation of energy.

Seven minerals, calcium, phosphorus, magnesium, potassium, sodium, chlorine and iron are needed by the body in the largest quantities. Another major mineral, sulphur, is available from the sulphur-containing amino acids. Fish is also a good source of other essential elements, but needed in much smaller quantities like manganese, copper, zinc, cobalt, arsenic, fluorine and iodine. Typical values of important minerals content in fish are presented in Table 1.1:

Table 1.1 : Typical Values of Minerals in Fish

Element	Content mg/100 g
Sodium	30-134.00
Potassium	19-502.00
Calcium	19-881.00
Magnesium	68-550.00
Iron	1-5.60
Chlorine	3-761.00
Iodine	0-2.73

1.1.5 Vitamins

Vitamins are low molecular weight substances performing important roles in regulating the bodily functions. These are substances, which the body cannot synthesise and hence must be supplied through food for normal growth and development. However, these are needed only in small quantities. Absence of a vitamin in the diet or its presence in insufficient quantities will lead to deficiency symptoms such as malaise or restricted growth in children.

Vitamins are divided into two groups : the fat-soluble and water-soluble vitamins. Fatty fish or fish with high content of liver oils have high concentrations of fat-soluble vitamins A, D and E. However, consumption of these vitamins in quantities larger than the needs of the body may be toxic and lead to metabolic disorders. Excessive consumption of vitamin D can cause problems like kidney stone. Fish is also a relatively good source of water-soluble vitamins of B group. The important vitamins found in fish are :

Fat-soluble

Vitamin A
Vitamin D
Vitamin E

Water-soluble

Riboflavin
Nicotinic acid
Pyridoxine (B_6)

Cyanocobalamin (B_{12})
Pantothenic acid
Biotin

Fish liver, apart from its content of vitamins A, D and E, is also a good source of B vitamins. Another important source of B vitamins is the fish roe.

1.1.6 Carbohydrates

Carbohydrate is not a nutritionally important component in fish. The major carbohydrate present in fish is glycogen which is a polymer of glucose and is stored mainly in the liver. Muscle of the live fish or crustacean generally contains between 0.1 and 1 per cent glycogen. But molluscs like clam or mussel have high glycogen content in the range 1-7 per cent. Glycogen also shows wide variations in molluscs.

Glycogen in fish declines rapidly during the stress and struggle associated with the method of capture; so also after death. Glycogen gets decomposed to small amounts of glucose, sugar phosphates and pyruvic acid and higher amounts of lactic acid. The formation of lactic acid and the consequent lowering of the pH has an important bearing on the rigor mortis in fish.

1.1.7 Non-protein nitrogenous compounds

The sarcoplasmic components, i.e. the fraction soluble in water and salt solutions of low ionic strength, contain, in addition to the proteins, some non-protein nitrogenous (NPN) compounds. 9-18% of the total nitrogen in the teleosts, 20-25% in molluscs and crustaceans and 30-35% in elasmobranchs belong to the NPN fraction. The major NPN compounds found in fish are free amino acids, peptides, creatine, betaines, nucleotides, quaternary ammonium compounds and, in some cases, some marine toxins. Urea is a major component of the non-protein nitrogen fraction in elasmobranchs. Some of these compounds are responsible for the smell and taste of fish. Similarly changes taking place in some of these compounds are indicative of the extent of spoilage of fish. These can be objectively assessed and are often measured as indices of quality. Occurrence and distribution of NPN compounds vary widely between different fishes and also within the same species depending on season, size, age, the stage of spawning etc.

1.1.7.1 *Free Amino Acids*

Free amino acids are present to the extent of 0.5-2% of the muscle weight of aquatic animals. They contribute to the osmoregulatory functions in the fish. They are also associated with the flavour of fish. Free amino acids get depleted in the fish during periods of starvation. Some of the free amino acids found in fish and shellfish are glycine, histidine and taurine. Shellfish like lobster, shrimp and crab, in general, contain more free amino acids than finfish and their flavour is attributed to this high content of free amino acids. Cultured fish are known to have less free amino acids than their wild counterparts.

Red muscle in the fish, particularly the scombroid fish like tuna, mackerel etc. contain more histidine than in the white muscle. Histidine, on decarboxylation by bacteria, gets deaminated producing histamine which is a toxic amine causing food poisoning.

1.1.7.2 *Nucleotides*

Adinosine triphosphate (ATP) is the main nucleotide present in fish at the time of death. ATP is degraded into hypoxanthine in the dead fish by the action of bacterial nucleotidases. The degradation takes place in steps, ATP ADP → AMP → IMP → Inosine → Hypoxanthine. Hypoxanthine is responsible for the bitter taste in spoiled fish. Some other nucleotides also may be seen in the fish muscle, but generally in insignificant quantities.

1.1.7.3 Peptides

Small amounts of peptides, carnosine, anserine etc., have been identified in the fish muscle extracts. Dark muscle tends to contain more peptides than the white muscle. Carnosine has some antioxidant properties.

1.1.7.4 Guanidino Compounds

The phosgines, creatine and creatine phosphate, serve as high energy reservoir and may be present in the muscle to the level of 300-700 mg%. White muscle contains more of this compound than the red muscle. Conversion of creatine phosphate to creatine occurs rapidly during stress and struggle.

1.1.7.5 Betaines

Betaines are abundant in molluscan and crustacean muscle and are found in the order of 400-1000 mg%. They contribute to the taste of fish. The most common is glycine betaine. Others include b-alanine betaine and homarine.

1.1.7.6 Marine Toxins

Some seafood may contain nitrogenous toxic compounds. Ciguatera and scombroid poisoning are the major ones. Others are saxitoxins, ciguatoxins, histamine and, tetradotoxin. The last two are produced by bacteria and others by marine algae.

1.1.7.7 Urea

Urea is present in all fishes, but only as a minor component, often less than 5 mg%, in the muscle of teleosts and shellfishes. However, it occurs in substantial quantities, to the extent of 2 per cent of the wet weight of fish, in elasmobranchs. Urea is involved in the osmoregulatory functions of the elasmobranchs. In dead fish urea is decomposed by the bacterial enzyme urease into ammonia and carbondioxide.

1.1.7.8 Quaternary Ammonium Compounds

Trimethylamine oxide (TMAO) is an important quaternary ammonium compound present in all marine fishes. It occurs to the extent of 1 per cent of the wet flesh weight in almost all marine animals. An essential distinctive factor between marine and freshwater fish is the total or near absence of TMAO in the latter. TMAO is considered one of the compounds contributing to the characteristic odour and flavour of marine fish.

Elasmobranchs have more TMAO in their muscle compared to the teleosts. Red muscle contains more TMAO than the white muscle. TMAO has a role in stabilising the intracellular proteins in the fish muscle. They are also known to have a role in the osmoregulation of fish.

Like urea, TMAO also undergoes bacterial degradation in dead fish producing trimethylamine. Content of trimethylamine in fish is an indicator of the extent of spoilage.

1.2 POST-MORTEM CHANGES IN THE FISH MUSCLE

Immediately after death the fish is soft and pliable and the texture is elastic to touch. After some time the muscle tissues will contract when the muscle becomes hard and stiff and the whole body becomes inflexible; the fish is said to be in rigor mortis. After lying stiff for some time, the fish gradually becomes soft and pliable again. The texture of the fish muscle is by no means the same in pre- and post-rigor fish. Post-rigor fish will retain pressure marks if pressed between fingers, whereas pre-rigor fish flesh being inelastic will not retain the impression. When the rigor is at its maximum the fish will be so stiff and inflexible that it cannot be bent. The *post-mortem* changes take place consequent to the following developments. Blood circulation ceases immediately on death resulting in the stoppage of supply of oxygen to the tissues. However, the native enzymes remain still active. These enzymes can cause degradation of the muscle components by autolysis because the tissues lose their defence mechanism against microorganisms. Absence of antioxidants due to the stoppage of circulation leads to oxidation of lipids present in the tissues by atmospheric oxygen and several other changes take place which can be summarised as follows.

1.2.1 Glycolysis

In the live fish, with the help of oxygen supplied through blood, the muscle oxidises glycogen to carbon dioxide and water producing chemical energy. After death, as the blood circulation stops, the muscle cells do not receive oxygen any more. Aerobic oxidation ceases; however, anaerobic oxidation of glucose with the help of enzymes present in the muscle takes place producing lactic acid. This process is called glycolysis.

Due to accumulation of lactic acid the pH of the muscle falls. Glycolysis will continue until the glycogen content is completely used up. Because of the reduction of the pH of the muscle to the acidic side the fish flesh gets protection against bacterial invasion.

1.2.2 Rigor Mortis

Rigor mortis is the phenomenon of stiffening of the muscle taking place shortly after death. During rigor mortis the muscles become hard and inelastic unable to be stretched out. Because of the lowering of the pH the proteins lose the water holding capacity, i.e. the proteins get progressively denatured. Further, the myofibrillar proteins actin and myosin combine to form actomyosin resulting in the muscle shrinkage causing stiffness of the muscle. The development of stiffening is also related to the degree of degradation of adenosinetriphosphate (ATP) in the muscle.

ATP, which is an energy donor in metabolic processes, is formed in the live fish by the reaction between adenosinediphosphate (ADP) and creatine phosphate. Creatine phosphate is a reservoir of energy in the muscle cell. On death this reaction stops and consequently the production of ATP. Further, the reserve of ATP in the tissues becomes rapidly degraded.

The combination of actin and myosin is accompanied by step-wise hydrolysis of ATP→ADP→Adinosinemono phosphate (AMP) → Inosine monophosphate (IMP) → Inosine → Hypoxanthine. The hypoxanthine content will progressively increase with lapse of time and its concentration is considered as an index of freshness of fish. The stiffness of the muscle begins when the content of ATP reaches a low level of about 5 per cent of the original.

IMP has distinct taste and is a flavour component in fish whereas hypoxanthine is bitter and imparts an off flavour.

Rigor mortis is of great significance in fish processing, particularly freezing. If the fish are filleted pre-rigor, the fillets pass into rigor before attaining the freezing temperature resulting in severe contraction of the muscle and causing heavy loss in weight on thawing.

1.2.3 Hydrolytic Changes

Lysosomes in the cell cytoplasm contain several hydrolases enzymes like proteinases, nucleotidases, lipases etc. which can degrade important tissue components like proteins, nucleic acids, fat, carbohydrates etc. The lysosomes are enclosed by a membrane which can be easily damaged. Due to lowering of pH after death the lysosomal membrane gets ruptured releasing the hydrolyases resulting in the rapid degradation of tissue constituents. The degradation of proteins results in raising the water holding capacity and pH. The proteins are converted to proteases, peptones, polypeptides and amino acids which together with the nucleotides, minerals and vitamins constitute an ideal medium for microbial proliferation. This favours the onset of rapid microbial spoilage. The extracellular bacterial enzymes degrade the tissues further with the concomitant rise in spoilage indices like trimethylamine, total volatile bases etc.

1.2.4 Changes in Fat

Fish tissues are supplied with antioxidants like α-tocopherol through the circulatory system and hence the fat in the living tissues does not become oxidised. However, with the stoppage of circulation on death, the fat in the tissues undergoes rapid oxidation by atmospheric oxygen. This is further catalysed by the 'heme' proteins present in the tissues. The rate of these changes will depend on the temperature of storage and access to oxygen.

1.3 CHANGES TAKING PLACE IN DIFFERENT TISSUE CONSTITUENTS DURING SPOILAGE

1.3.1 Water and Protein

Water in the tissues exists in two different forms, free water and bound water. Water molecules attached to other molecules like protein etc. through strong or weak chemical bonds constitute the bound water. The bound water may be an integral part of the molecule. Removal of water by any process will result in alterations of the structure of the molecule. In other words, when the structure of such molecule is altered by any process, the bound water is released.

The physical and biological properties of protein are due to the unique structure of these molecules. In a living system this structure is kept in tact. Changes in pH, temperature etc. will affect the structure of the protein molecules, especially once the system is dead. Under such conditions the natural form of protein is lost (denaturation of protein) which will result in loss of texture, reduced water holding capacity etc. Different methods of handling and processing affect these constituents in different ways.

1.3.2 Lipids

The important deterioration in the lipids are autoxidation and hydrolysis. Fish lipids are very reactive because of the presence of highly unsaturated fatty acids. Certain enzymes in the tissues act as pro-oxidants and play very important role in the oxidation and rancidity. Exposure to oxygen initiates the reaction and then it propagates by a free radical mechanism. Traces of iron, copper etc. present in the system act as catalysts.

Primary products of autoxidation viz., peroxides and hydroperoxides decompose rapidly into various products. Peroxide formation is slow in the initial stages, as in the frozen products. This period is called the induction period. The length of induction period depends on many factors like initial freshness of the fish, temperature of freezing and storage, biochemical composition of the tissues etc. After the induction period the peroxide concentration increases and reaches a maximum. A decrease in the peroxide concentration follows this phase. Therefore, peroxide concentration alone cannot serve as an indicator of oxidative rancidity. Fish stored for a longer time may have lower levels of peroxides than those stored for shorter periods. Adehydes, ketones etc. are formed as a result of decomposition of the peroxides. Most of these are volatile and contribute to the odour and flavour of fish. They interact with proteins and other tissue constituents forming complex molecules. Yellow or brownish discolouration occurs in frozen fish in extreme cases of oxidation. Estimation of the oxidation products or their secondary products cannot give a direct measure of the degree of oxidative rancidity because of the very complex nature of the formation, decomposition or interaction of these entities. The best way to assess the oxidative rancidity will be organoleptic evaluation.

Lipid hydrolysis is caused by the lipolytic enzymes present in the tissues. Accumulation of free fatty acids results in further deterioration in quality. Free fatty acids accelerate protein denaturation and, in their free state, are more susceptible to oxidation than in the esterified state. Therefore, hydrolysis of lipids indirectly promotes lipid oxidation.

1.4 SPOILAGE INDICES

In the trade, freshness is usually judged based entirely on the appearance, odour, and texture of the raw fish. Since this assessment depends on the senses, these factors are called sensory or organoleptic parameters. Careful and accurate descriptions of these properties for different stages of spoilage are available for many species of fish. In the raw fish, appearance of the eyes, flesh, odour and texture are the factors assessed. Odour, texture and flavour are the factors assessed in cooked fish. It is possible to determine the degree of freshness to a very satisfactory level based on the sensory characteristics. However, this is a subjective method and individual preferences may, to some extent, influence the judgement.

Measurement of the concentration of certain chemicals in the tissues serves as an objective method of assessing the quality. The changes in odour, flavour and texture of the fish flesh appear to run parallel to certain chemical changes in the spoiling fish. Therefore, determining the concentration of products of chemical and bacterial processes may be useful in determining the extent of spoilage. Although the method is objective, measurement of chemical indices alone does not provide satisfactory results always. Since the spoilage is the result of many complex changes and most chemical methods involve the measurement of only one, or at the most a few indices, no chemical method can give the same degree of accuracy in measuring freshness as does a properly trained taste panel. In spite of this, certain chemical indices provide valuable information about the quality of the product, and hence it is useful to determine them during the course of assessment of quality. Some of the important indices are detailed below.

1.4.1 Total Volatile Bases/Total Volatile Nitrogen (TVN)

Spoilage of fish is accompanied by the release of several volatile compounds like trimethylamine, ammonia etc. The concentration of these compounds in the tissues may indicate the degree of spoilage, particularly in the later stage of spoilage. The recommended upper acceptable limits vary from 30 to 60 mg/100g fish flesh. The wide variations in the levels of TVN are due to differences in the composition, bacterial flora, handling methods etc.

1.4.2 Trimethylamine (TMA)

Trimethylamine is one of the chemical indices that has the nearest correlation with sensory evaluation in many species of fish. Trimethylamine oxide (TMAO), a naturally present component in the marine fish, is broken down by bacteria to TMA, related chemically to ammonia. Very fresh fish has only very low content of TMA and its concentration increases as spoilage advances. It has been suggested that TMA is a product of early stages of spoilage and may be lost during storage. As in the case of TVN, the level of TMA also varies due to the influence of many factors.

1.4.3 α-amino Nitrogen

During spoilage the tissue proteins are degraded into simple peptides, amino acids etc., thereby increasing the content of free α-amino groups in the muscle. The estimation of α-amino nitrogen, thus, serves as an index of the extent of spoilage.

1.4.4 Ammonia

Ammonia is generally a product of protein decomposition. In some cases it has been found that the levels of ammonia serve as an indicator of spoilage. But in some other cases ammonia appears only at a relatively later stage of spoilage. Ammonia will appear as the result of decomposition of urea in elasmobranchs

1.4.5 Volatile Reducing Substances (VRS)

Compounds responsible for off odour and flavour are released during spoilage of fish. These are volatile compounds and many of them are reducing agents. Therefore, the quantity of volatile reducing substances in the tissue has been considered to be an indicator of the extent of spoilage. However, as in the case of other chemical indices, the results are not entirely satisfactory. Direct correlation between VRS and organoleptic assessment has not been possible in many cases.

1.4.6 Hypoxanthine

The enzymes present in the fish tissues decompose adenosinetriphosphate (ATP) through a series of steps to hypoxanthine. Formation of hypoxanthine in the tissues is somewhat proportional to the extent of spoilage during the early stages.

1.4.7 Peroxide Value (PV)

Oxidative rancidity is one of the most important factors that determine the acceptability of the fish during processing and storage. In fatty fish this is one of the most important quality problem. Peroxide value is a measure of the degree of oxidation of the fat. However, as discussed earlier, because of the very complex nature of the reactions involved correlation between peroxide value and organoleptically detectable rancidity is often highly variable.

1.4.8 Thiobarbituric Acid (TBA) Value

Oxidised fat reacts with thiobarbituric acid to produce a red pigment. The reactive material is supposed to be malonaldehyde mainly produced from methylene separated dienoic or polyenoic acids. TBA test is used to determine the quantity of malonaldehyde in the fish. However, direct correlation with organoleptic evaluation is possible only to a limited extent.

1.4.9 K-value

K-value index which relates the freshness to the autolytic degradation of nucleotides in the fish muscle is becoming popular as an index of freshness. The autolytic degradation products of the nucleotides are separated and collected by means of anion-exchange chromatography and measured using a spectrophotometer. The K-value is arrived at using the relation

$$\text{K-value (\%)} = \frac{HX+HXR}{ATP+ADP+AMP+IMP+HX+HXR} \times 100$$

Standards may have to be set for different species of fish and shellfish.

1.5 NUTRITIONAL QUALITIES AND SPOILAGE

The importance of fish as a food is primarily because of its unmatched nutritional qualities. Fish protein is one of the best from the nutritional pint of view. It is easily digestible and it contains all the essential amino acids. Fish proteins also have hypocholesterolemic properties. Similarly, fish lipids, apart from serving as a source of energy, also have special nutritional importance. The polyunsaturated fatty acids in the fish oils are believed to be cholesterol removing factors. Long chain n-3 polyunsaturated fatty acids, especially the eicosapentaenoic and docosahexaenoic acids, play important roles in the functioning of the nervous system and in the process of blood clotting.

During spoilage many of the nutritionally important compounds are destroyed and some toxic substances accumulate. Changes in the structure of proteins and the lipid molecules during spoilage may render them nutritionally not available. Aggregation or polymerisation of proteins may make them indigestible. Partial hydrolysis of protein releases certain amino acids preferentially and such losses result in decrease in the nutritional qualities. Oxidation of fatty acids molecules brings about structural changes in them and this not only affects the organoleptic acceptability, but also lowers the nutritional value of fish and in extreme cases makes it toxic.

Suggested Reading

Borgstrom, G. (ed.) (1965) *Fish as Food*. Vol. I. Academic Press Inc., New York and London

Clucas, I.J. and Ward, A.R. (eds) (1996). *Post-harvest Fisheries Development: A Guide to Handling, Preservation, Processing and Quality*. Chatham Maritime, Kent, ME4 4TB, United Kingdom.

Connell, J.J. (ed.) (1990). *Control of Fish Quality*. Blackwell Scientific Publications Ltd., Oxford, UK.

George, J. F. Jr. and Roy, E.M. (1992) *Advances in Seafood Biochemistry - Composition and Quality*. Papers of ACS Annual Meeting at New Orleans, Louisiana. Technomic Publishing Co. Inc., Pennsylvania

Gopakumar, K. (ed.) (1993). *Indian Food Fishes. Biochemical Composition*, Central Institute of Fisheries Technology, Cochin, India.

Heen, E. and Kreuzer, R. (1961) *Fish in Nutrition*. Fishing News (Books), London.

Martin, R.E., Flick, G.J., Hebard, C.E. and Ward, D. R. (1982). *The Chemistry and Biochemistry of Marine Products*. The AVI Publishing Company, Westport, Connecticut, USA.

Pomeranz, Y. and Meloan, C. E. (eds) (1978). *Food Analysis Theory and Practice*. The AVI Publishing Company, Westport, Connecticut, USA.

Sikorski, Z.E., Pan, B.S., Shahidi, F. (eds) (1994) *Seafood Proteins*. Chapman Hall, New York and London.

Suzuki, T. (ed.) (1981) *Fish and Krill Protein - Processing Technology*. Applied Science Publishers, London.

2
BACTERIOLOGY

MICROORGANISMS exist as two distinct types of cells called procaryotes and eucaryotes. Eucaryotic cells have a true nucleus with multiple chromosomes enclosed in a nuclear membrane and are the unit of structure in plants, animals etc. Procaryotic cells are smaller, less differentiated and the nucleus consists of a single, naked cyclic DNA molecule without a nuclear membrane. Bacteria are microscopic organisms consisting of a single procaryotic cell and are found widely distributed in nature. They are found in all environmental conditions and are present in the intestinal tracts and body surface of animals. They are present in soil and the natural waters of ponds, lakes, reservoirs, rivers and oceans.

Bacteria are both useful and harmful. They are involved in the decomposition of dead tissues of plants and animals. Bacteria are used in the industrial process like fermentation, in medicine etc. On the harmful side bacteria are associated with several diseases in man and animals. Such bacteria are called pathogens. Almost all foods of animal and vegetable origin are decomposed by bacteria unless they are preserved and protected against infection.

2.1 MORPHOLOGY

Bacteria may be classified into one or other group on the basis of their morphology and staining reactions. Morphological features of importance are size, shape, grouping of the cells and their possession of any distinctive structures like endospores, flagella, capsules and intracellular granules.

2.1.1 Shape

Bacteria exist in different shapes. The shape is provided by a strong, relatively rigid cell wall. Three basic shapes in which bacteria exist are spheroidal or coccoid, cylindrical or rod shaped and spiral or filamentous (Fig. 2.1).

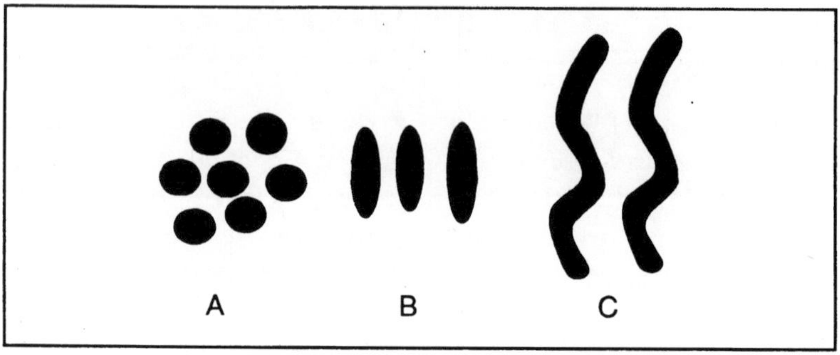

Fig. 2.1 : Morphological forms of bacteria (a) spherical, (b) cylindrical and (c) spiral

2.1.2 Size

The size of bacteria ranges from 0.5 to 100 μ in length and 0.5 to 2 μ in diameter (one μ (micron) = 1 ÷ 1000 of a mm). Spheroidal bacteria have diameter of 0.75 to 2.0 μ. A medium sized rod has a width of 0.5-1.0 μ and length about 3 μ. Some rods develop filamentous forms 100 μ or longer.

2.2 CYTOLOGY

The bacterial cell comprises of a rigid semi-permeable outer wall and a semi-permeable cytoplasmic membrane within enclosing a soft gel-like viscous watery solution called protoplasm. The protoplasm contains a variety of organic and inorganic solutes, a nucleus and numerous small particles called ribosomes.

2.2.1 Cell Wall

Bacterial cell wall is a strong, relatively rigid and semi-permeable structure about 10-25 μm thick. It has some elasticity and is responsible for maintaining the characteristic shape of the bacterium. Cell walls also support the weak plasma membrane against the internal osmotic pressure of the protoplasm. It consists of a meshwork formed by the arrangement of macromolecules of mucoproteins, polysaccharide chains and peptide bridges. The rigidity of the cell wall is contributed by this meshwork.

2.2.2 Gram stain

One of the methods of classification of bacteria depends on their reaction to Gram staining. If, after Gram staining, the bacteria retain the violet colour, they are said to be Gram-positive; if the violet colour is eluted and only red colour remains the organisms are Gram-negative.

2.2.2.1 Method of Gram Staining

A thin bacterial film on a glass slide is covered with a solution of crystal violet for one minute and is then washed off with water. A dilute solution of iodine is added over the same film and allowed to remain for one minute. After this the slide is washed with rectified spirit until no more colour is released from the film. Finally, a counter stain like safranine (red) is poured over the film and allowed to remain for half a minute. The slide is then washed with water and allowed to dry.

Some bacteria retain the violet stain even after washing with alcohol so that the red counter stain is not taken up by the stain. They appear violet under microscope and are termed Gram-positive. Others do not retain the violet stain, but take up the counter stain and appear red under microscope. They are Gram-negative. There are some bacteria which are either weakly Gram-positive or Gram variable.

There is considerable difference between the cell wall structures of the bacteria of Gram-positive and Gram-negative groups. Gram-positive bacteria have thicker cell walls 15-20 μm thick. It contains three or four amino acids, a small quantity of lipid material and large amounts of polysaccharides and techoic acids. Cell wall is thin, 10-15 μm, but chemically more complex in Gram-negative bacteria. It contains proteins, considerable amounts of lipid material and little polysaccharides, but no techoic acid.

Cell wall also plays an important role in cell division.

2.2.3 Cytoplasmic Membrane

All that lies within the bacterial cell is called the protoplast. Protoplast is contained within a very thin elastic cytoplasmic membrane, 0.5-10 μ thick, consisting mainly of lipoproteins. Its main function is active selection and transport of nutrient solutes into the cell wall and waste products out of it. The membrane itself has little mechanical strength and is supported by the external cell wall.

2.2.4 Protoplasm

Protoplasm is a viscous watery solution or soft gel containing a variety of organic and inorganic solutes. It is filled with numerous tiny granules, 10-30 μm in diameter, called ribosomes consisting chiefly of ribonucleic acid (RNA). RNA is the site of protein synthesis in the cell.

2.2.5 Nuclear material

Nuclear material consists of a large cyclic deoxyribonucleic acid (DNA) molecule containing genetic information essential to the bacterium. Multiple replication forks appear in fast growing cells enabling division of nuclear material to keep pace with the cellular growth.

2.2.6 Intracellular or Inclusion Granules

In many species of bacteria some round granules are found in their protoplasm. These, however, are not essential structures or may be even absent under certain conditions of growth. These are some excess of metabolites stored as a reserve nutrient and may consist of volution lipids, glycogen, polysaccharides and sulphur.

2.2.7 Flagella

Some rod-shaped and coccoid bacteria show hair-like appendages on their body called flagella. Flagellum confers motility to the bacteria, especially in a liquid medium, and acts as an organ of locomotion. Flagellum is a long thin filament in a helical form. It is about 0.2 μ thick and several times as long as the bacterial cell. Flagellum originates in the bacterial protoplasm and extends through the cell wall. One or more flagella may be seen on a cell depending on the species. When arranged round the whole cell, they are said to be peritrichous; and when on one or both ends, polar. Several polar flagella may aggregate into a bundle and function as a tail.

Flagellum consists largely or entirely of a protein, flagellin, similar to myosin. Motility of the bacteria helps them to take up nutrients by continuously changing the body fluids in contact with the bacterial cell structure.

2.2.8 Fimbriae or Pili

Certain Gram negative bacteria possess very fine hair-like appendages called fimbriae or pili. Like flagella they also arise from protoplasm and project through the cell wall; however, they are much shorter and more or less straight. They are arranged peritrichously and single cell may have up to 500 fimbriae. They have an adhesive property which enables them to attach to one another or to a substratum. In pathogens it helps them to attach to the specific host cells and erythrocytes.

2.2.9 Bacterial Capsule or Slime Layer

Some bacteria form a thick gelatinous covering layer or capsule outside their cell wall. Capsular gel or slime is water with about 2 per cent solids, usually a complex of polysaccharides, polypeptides or protein. Capsules play a significant role in determining the immunological specificity of bacteria and in conferring virulence.

Loose or free slime is produced extracellularly by many capsulate organisms and some non-capsulate species. It is a viscid colloidal material chemically similar to the capsules and secreted outside them.

The growth of capsule forming bacteria causes the type of spoilage called 'ropiness'.

2.2.10 Bacterial Spores

Some bacteria, particularly *Clostridium* and *Bacillus*, develop a resting phase called endospore. Sporulation takes place only when growth is arrested as a response to starvation or when some essential nutrients for vegetative growth are not in supply or under other adverse environmental conditions like heat, chemicals, desiccation etc. The spores are formed inside the protoplasm of the parent vegetative cell. It consists of a central core of condensed protoplasm containing the nuclear body and acquires a thick covering layer, the 'cortex' and a thin but tough outercoat, the 'spore coat'.

In 'sporulation' each vegetative cell forms only one spore that on subsequent 'germination' gives rise to a single vegetative cell.

Spores are more resistant to chemical and physical influences than are vegetative cells. They are sensitive to heat only above 100°C. Resistance is considered to be due to impermeability of cortex and spore coat, high content of calcium and dipicolenic acid, very low content of water and the consequent very low metabolic and enzymatic activity.

Spores germinate when external conditions become favourable, particularly with the availability of moisture and nutrients. Spores surviving heat is more exacting in their nutritional needs for germination like specially enriched culture media.

2.2.11 Chemical Composition of Bacteria

Seventy to Ninety per cent of the bacterial mass is water. Hydrogen, carbon, oxygen and nitrogen constitute 99 per cent of the mass. Only 12 other elements, sulphur, phosphorus, magnesium, calcium, sodium, potassium, chlorine, iron, manganese, copper, zinc and molybdenum, occur in all species of bacteria. Most of the carbon and nitrogen of the bacterial cell mass are contained in amino acids linked to proteins which constitute up to 50 per cent of the dry weight of the cell. Other macromolecules constituting the bacterial cell in the descending order are nucleic acids, carbohydrates and lipids. The four macromolecular types are usually found in combination with each other like lipoprotein, lipopolysaccharides, glycoproteins and nuclear proteins. Other constituents of bacterial cell may include alcohols, waxes, pigments and vitamins.

On dry weight basis bacteria may contain approximately 50 per cent carbon, 8 to 15 per cent nitrogen and 2 to 15 per cent ash.

2.2.12 Nutrition

Based on their nutritional needs bacteria are classified into two, autotrophs and heterotrophs. Autotrophic bacteria are those which can synthesis their protoplasm from simple inorganic compounds and are, therefore, considered not significant in food spoilage. Heterotrophs need organic materials as a source of carbon and energy. They cannot synthesise their complex protoplasm from simple inorganic salts. Their nutritional needs are complex and must have organic compounds such as proteins, peptones, amino acids, vitamins and several minerals. Pathogens causing diseases and saprophytes responsible for spoilage of food belong to this group.

Bacteria absorb nutrients through their cell. The nutrients must be available in an assimilable form for the bacteria to absorb them. It is the enzymes elaborated by the bacteria which accomplishes the task of breaking down complex nutrients to simpler compounds which can pass through the bacterial cell wall. The nutrients undergo further enzymatic changes within the cell. The end products resulting from these decompositions are hydrogen, carbon dioxide, alcohol, lactic acid and other acids from sugar and, amino acids, carbon dioxide, hydrogen sulphide, indole and other foul smelling compounds from protein.

2.3 BACTERIAL GROWTH

Growth of bacteria is profoundly influenced by the environmental conditions and availability of nutrients. The environmental conditions of significance are the presence or absence of oxygen, temperature and pH of the medium.

2.3.1 Oxygen Relationship

The principal gases that affect bacterial growth are oxygen and carbon dioxide. On the basis of influence of atmospheric oxygen on their development bacteria are grouped as follows :

- aerobic bacteria or aerobes, which require molecular oxygen for growth
- anaerobic bacteria or anaerobes, which are unable to grow in the presence of molecular oxygen
- facultative anaerobes, which can grow both in the presence and absence of molecular oxygen
- microaerophiles, which will grow under oxygen tension less than that in the air; or in other words bacteria that grow in the presence of very small quantities of molecular oxygen.

2.3.2 Effect of Temperature

Bacterial growth is significantly influenced by temperature. Each species of bacteria grow at a temperature within a certain range. Within this range there is an optimum temperature of growth, i.e., a temperature at which cell reproduction is most rapid. With reference to the temperature requirements for their growth, bacteria are commonly classified into three broad groups.

2.3.2.1 *Psychrophiles*

As the name indicates, psychrophiles are cold loving bacteria. They grow readily at temperatures between 0°C, or even slightly lower, and 20°C, and have an optimum growth temperature of about 15°C. Psychrophiles are considered to be of great importance in low temperature preservation of fish as they are capable of spoiling fish preserved at low temperatures. The psychrophilic bacteria primarily associated with spoilage in fish are prominent in the genera *Pseudomonas, Achromobacter* and *Flavobacterium.*

2.3.2.2 *Mesophiles*

These form a group of bacteria that show maximum growth between 20 and 45°C. The optimum temperature of growth lics between 30 and 37°C and do not develop below 5°C. Majority of the bacteria belongs to this group; particularly the disease causing or pathogenic bacteria. Mesophiles generally consist of terrestrial flora like *Bacillus* anaerobic types, *Coryneforms* and *Micrococcus.*

2.3.2.3 *Thermophiles*

Thermophiles grow best between 45 and 65°C. They have an optimum growth temperature of about 55°C. Some of the bacteria occurring in natural hot springs can survive at a temperature of even 85 to 90°C.

Temperatures outside the range for growth of different groups have a lethal effect on the concerned bacteria.

The range of temperature of maximum and optimum growth of bacteria can be summarised as in Table 2.1.

Table 2.1 : The Range of Temperature of Maximum and Optimum Growth of Bacteria

Type of bacteria	Growth temperature °C	Temp. for optimum growth °C
Psycrophile	0-20	15
Mesophile	20-45	30-37
Thermophile	45-90	55

2.3.3 Effect of pH

Acidic or alkaline pH are harmful to the growth of bacteria. The optimum pH of the bacterial growth medium is usually between 6.5 and 7.5. pH beyond this, either in acidic or alkaline, range is unfavourable. Only some special type of bacteria and fungi require pH either on acidic or alkaline side for their growth.

2.3.4 Effect of Salt

Most bacteria require a low concentration of salt for their growth. Marine bacteria require around 0.5–3.0% salt. Many bacteria do not grow when the salt concentration is in the range 10·15%. However, halophilic bacteria can develop even in concentrated solutions of sodium chloride.

2.4 REPRODUCTION AND GROWTH

2.4.1 Reproduction

When introduced into a suitable medium or substrate under ideal conditions, bacteria increase rapidly in number within a relatively short time. Bacterial multiplication takes place by a process called binary fission or transverse fission. The bacterial cell grows in size, usually elongating to twice its original length. The protoplast becomes divided into two approximately equal parts by the ingrowth of a transverse septum from the cell wall. In some species the cell wall septum splits into two and the daughter cells separate almost immediately. In some others the cell walls of the daughter cells remain continuous for a while and the organisms grow adhering in pairs, clusters, chains or filaments. Under favourable conditions, growth and division takes place rapidly, almost every half hour or less. The new cells undergo the same phases and bacterial multiplication continues (Fig. 2.2).

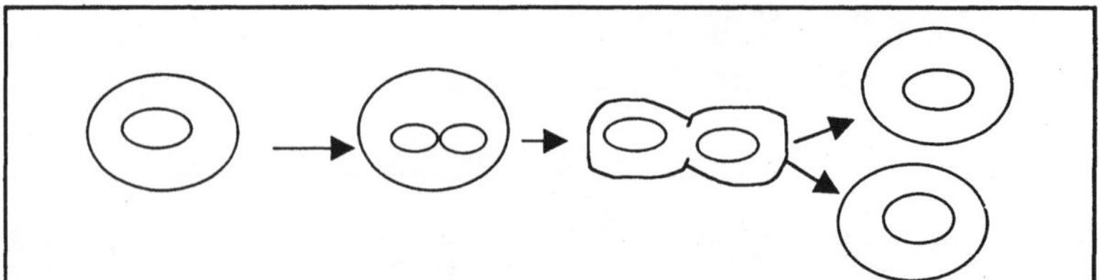

Fig. 2.2 : Bacterial multiplication

2.4.2 Growth

Rate of growth varies with different bacteria. It is also affected by such factors as age of culture, temperature, available nutrients, genetic make up of the organism etc. Maximal growth takes place at the optimum temperature.

Four distinct phases have been identified in the bacterial growth pattern. When a few bacterial cells are introduced into a fresh medium and incubated at a suitable temperature, the growth of bacteria follows the pattern shown in Fig. 2.3.

2.4.2.1 The Lag Phase

Immediately after inoculation the bacteria takes some time to adapt to the environment e.g. if the nutrients are different in the new environment from those in the previous environment in which the cells were growing, they may have to synthesise new enzymes to break them down. The organisms will be active metabolically, but little or no cell division takes place at this phase. This is called the lag phase. Lag phase may last few hours.

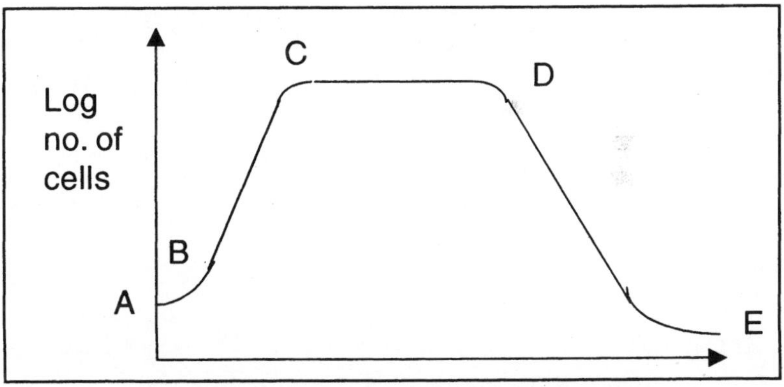

Fig. 2.3 : Bacterial growth pattern AB lag phase, BC logarithmic phase, CD stationary phase and DE death phase

2.4:2.2 The Logarithmic or Exponential Phase

This is the period at which the cells divide at a constant rate and hence is the period of most rapid growth. There exists a linear relationship between the logarithm of the numbers of bacteria and time. Under most favorable conditions cells divide in about 10 minutes. Cells in the log phase have a very high rate of metabolism and are very sensitive to heat and antimicrobial agents. Duration of the log phase varies with species and is influenced by temperature and other environmental factors. The log phase lasts few hours.

2.4.2.3 *The Stationary Phase*

At the end of the logarithmic phase the rate of cell division becomes reduced and for a period the number of cells remains constant. This happens because the number of cells produced is just sufficient to balance the number of cells dying, thus keeping the number of viable cells stationary. The slowing down of growth is attributed to exhaustion of essential nutrients in the medium and or accumulation of toxic waste products.

2.4.2.4 *The Death Phase or The Phase of Decline*

After sometime in the stationary phase the cell division ceases and cells begin to die in large numbers. Cause of death is the same as that in the stationary phase. Death can be defined as loss of ability to grow and multiply when the cells are transferred into a fresh nutrient medium. In some cases there may be a rapid fall in the total count. This happens if the dead cells undergo autolysis in which cells are broken down by their own cellular enzymes.

The pattern of bacterial growth under ideal conditions is presented above. In practice, however, this is need not be followed every time because of the possible variations in the environmental conditions.

2.5 BACTERIAL TOXINS

Food-poisoning bacteria produce toxins, the substances causing diseases as a result of their injurious effects on the body tissues. Toxins are usually antigenic; they stimulate the body tissues to produce various specific neutralising substances collectively called antibodies. Bacterial toxins are of two types, the exotoxins and endotoxins

2.5.1 Exotoxins

Exotoxins are extracellular. They diffuse freely into the medium in which the bacteria grow and can be easily separated from the cells by filtration. They are very active even in small quantities and stimulate the formation of antitoxin type antibodies which specifically neutralise the toxin, and renders it harmless. Symptoms caused by the toxins are highly characteristic and are usually recognisable by the effect they produce. Exotoxins are inactivated at 60°C in an hour. Typical among the exotoxins are those produced by botulism, tetanus and diphtheria organisms.

2.5.2 Endotoxins

Endotoxin is an integral part of the cell and hence cannot be separated from them by filtration. Symptoms produced by endotoxins in the body are less characteristic than those produced by exotoxins and can be produced only by the actual cells or cell extracts. Endotoxins do not generally stimulate antitoxin formation. The antibodies produced are antibacterial types which directly destroy the bacteria or the agglutinins that mobilise them to aggregate into clumps. Endotoxins are relatively heat-stable.

2.6 BACTERIOLOGY OF FISH AND ENVIRONS

Fish, irrespective of the source from which it is caught, will contain microflora, the nature of which will depend on the microflora of the water from which they are caught. The type and abundance of bacteria in water will vary widely depending on whether it is the open sea, brackish water or fresh water. Each system will support its own characteristic types of bacteria, which in turn will influence the bacterial quality of fish caught from these respective waters.

2.6.1 Bacterial Quality of Water Bodies

2.6.1.1 Seawater

Seawater contains varied types of microflora. These are not free-swimmers, but are usually found associated with plankton and other marine flora and fauna. Far out in the open sea the flora is more or less typically marine. However, near shore, harbour mouths, estuarine areas etc., the water will always contain varying proportions of terrestrial bacteria. Marine bacteria are able to tolerate higher concentrations of sodium chloride than the terrestrial bacteria.

2.6.1.2 Brackishwater

Brackishwater areas are generally those where intermingling of fresh water and saline water takes place and the water will remain saline, though to a less extent than the seawater. The salinity will not be consistent as in the open sea, but will fluctuate with season, tidal influx etc. Brackish water, therefore, will support varying proportions of both freshwater and marine bacteria.

2.6.1.3 Freshwater – Lakes, Rivers

Freshwater also supports bacteria. Their principal characteristic is low tolerance to sodium chloride. However, river and lake waters are mostly near human habitations and hence are often open to faecal contamination. The principal danger is its contamination with pathogenic bacteria. Therefore, the fish caught from these sources are likely to be contaminated with disease causing bacteria.

2.6.2 Bacteria in Fish

The flesh and internal organs like liver, heart etc. of a healthy live fish are sterile. However, there are a few areas where substantial harbouring of bacteria can occur in a freely swimming fish. These are the intestines, especially when it contains food, the gills, the skin and, other exposed areas. As long as the fish is live, the activities of the bacteria are under check. They do not act on the fish flesh. However, once the fish dies the bacteria start attacking the fish flesh initiating spoilage. Fish flesh containing 100 million (10^8) bacteria is considered rotten and unsuitable as food

Bacterial load means the number of bacteria present in a unit mass or unit area. For example, the bacterial load in different parts of a freshly caught marine fish can be of the following order :

Skin	10^3–10^5/cm^2
Gills	10^5 – 10^6/g
Intestine	10^5 – 10^8/g

Usually the highest bacterial load is met with in the intestine followed by gills and, the lowest on the skin.

2.6.3 Bacterial Genera Associated with Fish

2.6.3.1 *Marine Fish*

Freshly caught marine fish have their microflora consisting mostly of Gram-negative rods. Their extent in the total microflora may exceed 90 per cent and at times the entire flora is constituted by this group. Presence of Gram-positive organisms is often quite insignificant. However, when the fish is landed on deck or on sorting and transport, a great deal of contamination with terrestrial organisms takes place and the extent of Gram-positive organisms shows sharp increase.

The predominant genera in the gills, intestines and slime on the skin in many freshly caught marine fish have been found to consist of *Pseudomonas* and *Achromobacter* which together accounts for 45-60 per cent of the total isolates. *Flavobacterium, Corynebacterium* and Micrococcus together form the next predominant group. The remainder consists of miscellaneous genera like *Bacillus, Aeromonas,* Photobacterium and the like.

2.6.3.2 *Freshwater Fish*

Freshwater fish, as can be expected, carry freshwater bacteria. The guts of freshwater fish carry bacterial species of the genera *Achromobacter, Pseudmonas, Flavobacterium, Vibrio, Bacillus, Clostridium* etc. However, these are flora associated with the guts of marine fish as well. In addition to the genera associated with marine fish species, *Aeromonas, Lactobacillus,* Streptococcus etc. are also seen in freshwater fish.

2.6.4 Microbial Spoilage

Bacterial spoilage essentially involves partial digestion of fish tissues by the bacteria in their process of growth and reproduction. Nutritional requirements of microbes and man are strikingly similar. The most nourishing of human foods is almost always ideal as food for bacteria. That accounts for why highly nutritious foods like milk, egg, meat, fish etc. spoil rapidly when they become contaminated with bacteria in sufficient numbers. During the digestion process varying amounts of volatile and non-volatile compounds are liberated from the fish. Many products of protein degradation give out bad odour indicating that spoilage has set in. Volatile amines and organo-sulphurous compounds are responsible for this.

There are three main routes through which bacteria penetrate from the outer and inner surface into the fish flesh. These are :

- gills to blood vessels to flesh
- intestines to body walls to flesh
- skin (slime) to flesh

Bacteria may enter the flesh through any bruise caused during fishing or subsequent handling.

Gills are soft, moist and rich in blood forming an ideal environment for bacterial growth. Bacteria move rapidly from the gills to the flesh via the blood vessels. This underscores the importance of removing the gills and bleeding in fresh fish handling.

With lot of food in the stomach and intestines or after the lapse of some time after death, the bacteria will be active in these points and gain entry through the peritoneum into the flesh. This is important in handling fresh fish where evisceration is recommended before icing or any other type of preservation.

The slime on the outer surface of live fish acts as a barrier against bacterial invasion, because it may contain antibodies against bacteria and also some antibiotic substances. On death this defence mechanism ceases to work because those present are destroyed and new production ceases. Slime is an ideal medium for the growth of bacteria. The bacteria in the slime produce odoriferous substances that penetrate into the flesh.

Once the fish dies its natural defence mechanism fails and bacteria become active. Bacteria secrete digestive juices called enzymes. Bacteria and enzyme invade the tissues through gills, blood vessels, directly through the skin and through the lining of the belly cavity.

Whole round fish are attacked by bacteria from inner and outer sources. The speed of decomposition depends on the type of the fish, area from where caught, season, handling after catch and feeding status.

Bacteria and enzymes break down and dissolve the tissues they attack. Flesh of fish contains considerable amounts of non-protein nitrogen compounds. The enzymes naturally present in the fish produce autolytic changes, which increase the supply of nitrogenous foods like amines and amino acids, and glucose for bacterial growth. The bacteria converts these compounds to their break down products, the end product being hydrogen sulphide, mercaptans, indole etc. which are indicative of putrefaction.

2.6.5 Action of Bacteria on the Chemical Components of Fish

2.6.5.1 Degradation of Amino Acids

The bacteria in the fish first utilise the amino acids present in the fish flesh. The reactions involve decarboxylation and deamination.

• DECARBOXYLATION

Decarboxylation is the reaction involving the removal of the carboxyl group from the amino acids with the aid of enzyme *amino acid decarboxylase* producing a primary amine and carbon dioxide.

$$R\text{-}CH(NH)_2\ \text{-}COOH \rightarrow R\text{-}CH_2\text{-}NH_2\text{+}CO_2$$

This results in the accumulation of primary amines in the putrefying fish flesh. An example is conversion of lysine into cadaverine.

Scombroid fishes like tuna and mackerel have high contents of histidine in their muscle. Histidine decarboxylating enzymes acting on histidine produces histamine which causes food poisoning.

• DEAMINATION

Bacteria attack amino acids in another way which involves deamination. Ammonia is the product of deamination of amino acids. This causes strong ammoniacal odour in the spoiled fish.

2.6.5.2 *Proteolysis*

Bacterial degradation of proteins takes place more slowly than the degradation of low molecular weight substances such as polypeptides, peptones or even amino acids.

2.6.5.3 *Reduction of Trimethylamine Oxide*

Fresh bony marine fish contains 0.1-0.5% Trimethylamine oxide (TMAO) in their muscles which is reduced to trimethylamine by the bacteria :

$$(CH_3)_3\ N\text{-}O + 2H \rightarrow (CH_3)_3\ N + H_2O$$

TMA contributes to the typical smell of marine fish.

2.6.5.4 *Action on Urea*

Urea in the flesh of elasmobranchs is converted to ammonia by bacteria :

$$(H_2N)_2\text{-}C\text{=}O + H_2O \rightarrow NH_3 + CO_2$$

2.6.5.5 *Bacterial Degradation of Bulk Stored Fish*

When the fish is stored in bulk the bacteria use up the available oxygen and the environment becomes anaerobic. Facultative anaerobic bacteria develop and they degrade the sugar components to acetate and carbon dioxide and TMAO is simultaneously reduced to TMA. Under the anaerobic conditions thus developed amino acids get decomposed to ammonia. In this phase there is development of amino acid-degrading and ammonia-producing bacteria.

2.6.5.6 *Microbial Rancidification of Fat*

Certain microorganisms contain the enzyme lipoxidase which activates unsaturated fatty acid chains by releasing oxygen and resulting in the formation of aldehydes and ketones. Marine fish has high content of unsaturated fatty acids and hence are more prone to rancidity by bacterial action.

2.7 METHODS OF PREVENTION/CONTROL OF BACTERIAL SPOILAGE

The methods employed for prevention or control of bacterial growth are based on the characteristics of the bacteria like their water requirement to perform normal biological functions, sensitivity to heat, chemical preservatives and the like.

2.7.1 Icing

The simplest method of control of spoilage in fish is by icing. At or near the temperature of melting ice many bacteria become inactive and the onset of spoilage becomes delayed. In tropical conditions the iced fish will remain in good condition for a few days. However, continued storage in ice will result in an increase in the psycrophiles which can thrive at the temperature of iced storage and consequently spoil the fish.

2.7.2 Freezing

Freezing is a long-term method of preserving fish. Freezing binds water into ice crystals making it not available for bacterial action. During freezing the temperature of fish is brought down to sub-zero levels of –20 to –40°C and then stored at similar temperature. Freezing reduces the initial bacterial load by 60-90%. Freezing causes death or considerable reduction in the viability of several bacteria. At sub-zero temperatures the activities of remaining bacteria also come to a stand still and hence the fish keeps well over very long periods.

2.7.3 Drying

Processes like drying, salting smoking etc. involves the principle of reducing the water activity of fish to a level at which they cannot survive. Water activity can be reduced simply by removing a considerable portion of water by evaporation by some means or reducing the availability of water for the biological functions of the bacteria like salting or similar methods. Drying should ensure removal of water to the required extent, if not the product would spoil quickly.

2.7.4 Heat Processing

Bacteria can be destroyed or their count reduced in foods by subjecting the food to heat treatment. This method is employed in the canning preservation of fish. At the normal temperature employed in canning most of the bacteria are destroyed; however, the aim of canning is not to achieve absolute sterility in the product; but only commercial sterility. This implies that the heat treatment should be sufficient to destroy all pathogens present and the remaining ones should remain dormant and should not multiply in the can. Properly processed canned fish should keep well for a minimum of two years.

Pasteurisation is another method that brings about partial destruction of bacteria by application of heat. The reduction in the number of bacteria slows down the development of spoilage flora and delays spoilage.

2.7.5 Use of Antibiotics

The broad spectrum antibiotics, tetracyclines, particularly chlortetracycline (Aureomycin), oxytetracycline (Terramycin) and chloramphenicol have been used to extend the shelf life of fish. These are effective both against Gram-positive and Gram-negative bacteria of different types; however, have no effect on yeast and moulds. The tetracycline residues are destroyed by heat and, therefore, were considered safe in foods consumed after heating. However, it is possible that their use may induce resistance in microorganisms against them when used clinically in human or animal medicine.

In treating fish with antibiotics, the best results are obtained when the fish is treated immediately after catch. Therefore, convenient and most effective methods are by applications by means of the ice or chilled water in which the fish is held. Ice containing 1 ppm CTC has been found to bring about substantial improvement in keeping quality when compared to keeping in normal ice. Dipping in a solution containing the antibiotics and further storing in ice also is an acceptable method for extension of shelf life of many fish and shellfish.

2.8 BACTERIAL HAZARDS IN FISH

The natural flora on fish or shellfish caught from waters away from landmasses consists of bacteria which are harmless to human beings. However, during handling after catch fish get contaminated with varying kinds of terrestrial bacteria, many of which are of great sanitary significance because of their pathogenicity in humans. The chances of contamination with such bacteria are very high unless a very high degree of sanitation is maintained from the moment of catch through various handling, processing, storage and transport practices until the fish reaches the consumer. Knowledge about the pathogenic bacteria which are likely to be present in fish, their origin, pathogenicity, control measures etc. is very essential in hygienic handling preservation and processing of fish.

2.8.1 Bacteria of Public Health Significance

The bacteria of public health significance encountered in fish and their primary habitat are shown in Table 2.2.

Table 2.2 : The Bacteria and Their Habitat

Name of bacteria	Primary habitat
Escherichia coli	Intestine of man
Faecal streptococcus	Intestine of man and animals
Staphylococcus aureus	Man
Salmonella	Intestine of man and animals
Shigella	Intestine of man
Vibrio cholerae	Intestine of man
V. parahaemolyticus	Marine environment
Clostridium botulinum	Soil, marine environment
Cl. welchii	Soil, dust, intestine of man
Listeria monocytogenes	Natural environment

Many of these bacteria are of human origin. The occurrence of Coliforms, *E. coli* and faecal *Staphylococcus* is indicative of faecal contamination. While the presence of few numbers of these bacteria is permissible in fish and fishery products, presence of some others like salmonella is not permitted to any extent.

2.8.1.1 E. coli

E. coli are Gram-negative rod-shaped non-spore forming bacteria. The primary habitat of *E. coli* is the intestinal tract of humans. Therefore, their presence in fish is considered as an indicator of its faecal contamination. Offshore water generally does not contain any *E. coli* whereas they have been found in near shore water. The near shore waters, the deck of fishing boats, utensils etc. infected with *E. coli* also are sources of contamination.

E. coli is very sensitive to freezing and frozen storage. About 95% reduction of *E. coli* takes place during freezing at –40°C and complete elimination takes place during subsequent cold storage at –23°C for about three months.

Avoiding fishing from near-shore waters, maintenance of proper hygiene, sanitation and disinfection of boat deck, handling equipment etc. are the preventive measures to be adopted.

2.8.1.2 Faecal Streptococcus

Faecal streptococcus is a Gram-positive non-spore forming non-motile coccoid shaped bacteria. Its primary habitat is the intestinal tract of man and animals. As in the case of *E. coli* faecal streptococcus also is absent in offshore areas but are present in considerable numbers in near shore waters. Unclean boat deck, utensils, water and ice are the major contaminants besides the water of the habitat.

Faecal streptococcus is comparatively resistant to many adverse conditions. About 30% reduction takes place during freezing at –40°C; whereas not much further reduction takes place during subsequent storage at –23°C even beyond two years. Because of this higher resistance to sub-zero temperatures Streptococcus is considered a better indicator of the level of sanitation. The preventive measures recommended are the same as for *E. coli.*

2.8.1.3 Coagulase Positive Staphylococcus

This is a very important bacterium from the public health point of view. The organism, when grows in considerable numbers in foods, secretes an exotoxin which can cause serious intestinal disorders like gastro-enteritis. Coagulase positive staphylococcus is a Gram-positive, non-spore forming, non-motile coccoid bacteria.

The primary habitat of staphylococcus is man and is present in the sweat, ear gum, tear, throat, ulcers, carbuncles, boils and to a great extent in the post-nasal drips of persons suffering from cold. Almost 30 per cent of humans are carriers of *Staphylococcus aureus.*

Fresh caught fish does not contain any coagulase positive staphylococcus but contamination can take place onboard through the handlers. Its incidence in fish is indicative of poor personal hygiene of fish handlers.

Staphylococcus food poisoning is caused only by a few well-defined strains of *Staphylococcus aureus* and such strains are known to produce coagulase and hence called coagulase-positive staphylococcus.

In very small numbers staphylococcus is considered harmless, but can vigorously multiply and produce toxin due to careless handling at or near room temperature. The toxin produced is heat stable at 100°C and therefore normal cooking emperature will not eliminate it, even though the bacteria will be killed at this temperature. Nausea, vomiting, abdominal pain, diarrhoea, abnormal blood pressure etc. are the symptoms associated with staphylococcus food poisoning.

Staphylococcus aureus are partially eliminated during freezing at –40°C. There occurs a gradual reduction in their count during storage at –23°C.

Observance of strict personal hygiene of fish handlers, maintenance of cool temperature, say 5°C or less, during handling and processing etc. are the suggested preventive measures.

2.8.1.4 Salmonella

Salmonella produce enteric fever and gastro-enteritis. Infants and, elderly and under-nourished persons are more susceptible to salmonella infection and in such people disease is known to occur even from one single cell of salmonella. The symptoms are nausea, vomiting, abdominal pain and restlessness. The symptoms are more severe in the case of enteric fever. The onset of symptoms takes place within 12-24 hours of consuming the infected food.

The primary habitat of the bacteria is the intestine of man and animals. Polluted water from which the fish is caught, use of contaminated water for washing and processing, sorting fish on contaminated beach, handling fish by carriers of salmonella, inadequate personal hygiene, flies, rodents etc are the main sources of contamination of fish with salmonella.

Salmonella survive freezing temperature of –40°C and storage at –23°C. Therefore utmost care is needed to prevent contamination of fish with this organism. The following measures are generally suggested :

- avoid fishing in polluted waters
- avoid of near-shore water for washing
- avoid sorting the catch on beaches
- use only chlorinated water (10 ppm) in processing, ice making etc
- ensure strict hygiene of fish handlers, and in premises, utensils, equipment etc.
- avoid potential carriers from handling fish

2.8.1.5 *Shigella*

Shigella, belonging to the *Enterobacteriaceae* family is a bacterium often associated with intestinal diseases like bacillus dysentery. Man is the only known natural host for Shigella. This is a Gram-negative, non-motile, non-spore-forming rod. The symptoms of the disease caused by Shigella are abdominal pain, loose motion with blood and mucus and fever. The organisms pass the acid barrier of the intestine, multiply in the gut and produce ulceration of the large intestine and dysentery. The organism is more resistant to external conditions than any other enteric organisms. However, Shigella does not survive at temperatures above 60°C.

Fish gets infected with Shigella through contaminated water, ice, contact surfaces, fish handlers who are carriers of the organisms etc. The control measures suggested include sanitary and hygienic handling practices, good personal hygiene of fish handlers, temperature control etc.

2.8.1.6 *Vibrio cholerae*

Vibrio cholerae, like several other pathogenic bacteria, have their primary habitat in the human intestine. This is a Gram-negative comma-shaped aerobic spore-forming motile rod. These organisms multiply rapidly in the intestines of the victims and produce a toxin which acts upon the mucosal cells of the small bowel causing them to secrete large quantities of isotonic fluid at a rate faster than the colon can absorb resulting in watery diarrhoea. The clinical symptoms include sudden onset of nausea, vomiting and profuse diarrhoea with abdominal cramps. There will be rapid loss of electrolytes leading to profound dehydration.

2.8.1.7 *Vibrio parahaemolyticus*

Vibrio parahaemolyticus is a purely marine pathogen present in seawater and sea mud. It produces food poisoning when present in large numbers. The symptoms that include abdominal pain, vomiting, diarrhoea and fever become manifest within 12 hours of consuming the infected food. Most of the bacteria die during freezing at –40°C and complete destruction takes place during subsequent storage at –20°C within 10 days. Icing the fish immediately after catch and maintenance of strict hygiene are the recommended remedial measures.

2.8.1.8 *Clostridium botulinum*

Clostridium botulinum is an anaerobic Gram-positive spore-forming rod. The spores are highly heat-resistant. They produce a deadly toxin and the resulting food poisoning is called botulism. Symptoms develop within 12-24 hours of consuming the suspected food. Symptoms include nausea, vomiting, fatigue, headache, paralysis of the muscles, difficulty to talk, double vision, sound in the ears etc. Death occurs due to respiratory failure. Botulism is a problem in canned foods, which are not properly sterilised.

2.8.1.9 Listeria monocytogenes

Listeria monocytogenes is a Gram-positive non-spore forming motile rod widely distributed in the natural environment. It is a pathogen that can survive refrigeration, freezing and pasteurisation temperatures and is relatively resistant to common salt. It produces a dangerous condition called listeriosis which can cause abortion or still birth in pregnant women and cause several other diseases like meningitis, skin lesions, conjunctivitis, pneumonia/typhoid, spinal or brain abscess.

The discussions above indicate the necessity of maintaining a high degree of sanitation from the moment the fish is caught till it reaches the consumer's table. The important requirements are :

- use of good quality water at every stage of handling and preservation
- use of adequate quantity of good quality ice for preserving raw fish
- proper methods of storing fish
- regular cleaning of boat deck, fish boxes, fish holds, utensils, table tops, handling equipment etc. with good detergent and subsequent washing with chlorinated water
- maintenance of good personal hygiene of fish handlers.

Suggested Reading

Chichester, C.D. and Graham, H.D. (1973) *Microbial Safety of Fishery Products*. Academic Press, New York and London.

Doyle, M.P. and Decker. M. (1989) *Food-borne Bacterial Pathogens*. Marcel Dekker, New York.

Gunslaus, I.G. and Stanier, R.Y. (eds) (1960) *The Bacteria - A Treatise on Structure and Function. Vol. I. Structure.* Academic Press, New York and London.

Hayes, P.R. (1985) *Food Microbiology and Hygiene.* Elsevier Applied Science Publishers, New York and London.

Kruickshank, R. (1962) *Handbook of Bacteriology.* E & S Livingstone Ltd., Edinburgh & London.

Skinner, F.A. and Carr, J.G. (1976) *Microbiology in Agriculture, Fisheries and Food,* Academic Press, New York and London.

Thimann, K.V. (1955) *The Life of Bacteria - Their Growth, Metabolism and Relationships.* The Macmillan Company, New York.

Ward, D.R. and Hackney, C. (1991) *Microbiology of Marine Food Products.* The AVI Publishing Co., Westport, Connecticut, USA.

3
ONBOARD HANDLING AND PRESERVATION

FISH is a highly perishable commodity. Various changes take place in the fish the moment it is taken out of water leading to its spoilage ultimately making it inedible. Spoilage is brought about by the action of bacteria, enzymes as also due to autoxidation of the fat. The flesh of live fish is sterile, but various bacteria will be present in the intestine, gills and surface slime which become actively associated with the *post-mortem* changes accelerating spoilage. The type of flora and the extent of contamination with them will depend on the bacterial quality of the waters from which the fish is caught as also the sanitary conditions under which it is handled and preserved onboard. The fish caught may be carrying feed in their digestive tract and the powerful digestive enzymes which continue to be active even after death may attack belly walls and cause discolouration or 'belly burns' besides causing other proteolytic changes in the fish. The live tissues are supplied with antioxidants through the circulatory system. This fails on the death of fish and the lipids become oxidised by atmospheric oxygen giving rise to oxidative rancidity.

3.1 NEED FOR HYGIENIC HANDLING

Spoiled fish can be considered as fish that has become less acceptable to the consumer or is even totally inedible as a result of various physical, chemical and other *post-mortem* changes taking place in them. Fish may be distributed fresh or may be processed into different products and marketed.

It is very essential that the fish reach the consumer in a highly acceptable condition. This is still more important if the fish is intended for further processing in any style, because the quality and acceptability of the end product depend very much on the quality of the raw material. In either case the quality of the product reaching the end user will greatly depend on how the fish was handled onboard the vessel, how it was preserved, packaged, transported etc. Eternal alertness and extreme care are needed in handling, preserving and storing fish onboard. Primary responsibility for ensuring the quality of landed fish rests with those who handle it onboard.

3.2 HANDLING FISH ONBOARD

The main factors affecting the quality of fish onboard are

- cleanliness of the deck and fish holds
- quality and quantity of water used
- temperature at which fish is maintained
- the general handling practices adopted
- cleanliness of the equipment and utensils used in handling, packaging and storage
- personal hygiene of the fish handlers.

Good handling practices at sea should ensure that the fish retains its natural freshness to the maximum possible extent. The important requirements are cleaning the fish from dirt and other debris, chilling it immediately to prevent its temperature from rising and maintaining high standards of cleanliness including personnel hygiene during handling and storage.

3.2.1 Washing

Immediately after unloading the catch should be washed well in order to free it from dirt and other extraneous matters. Seawater, when taken away from shore, will be sufficiently clean and will have only low bacterial load and, therefore, will be quite good for washing fish. Water from near shore is likely to be contaminated with bacteria and other pollutants like industrial effluents and even faecal matters. Even when taken from distant open sea, water may be chlorinated to 10 ppm available chlorine level to make it safe for use. Washing cleanses the fish of most of the surface bacteria which, otherwise, can bring about early spoilage in fish.

3.2.2 Sorting

Depending on the gear employed, the ground fished etc. the composition of fish will vary widely. In pelagic shoals the catch will consist mostly of a single species whereas in trawl catch it will comprise of several species and of varying sizes, very small to very big. Therefore, after washing the fish must be sorted species-wise as also size-wise. Fish which are considered unfit for preservation or for human consumption by way of being bruised, diseased, decomposed or otherwise damaged shall be separated and thrown back to the sea or kept separately for probable conversion into fish meal or manure later. High value fish are generally carefully sorted out and suitably preserved. Often the catch is sorted out species-wise and-size wise and stored separately in separate containers.

3.2.3 Evisceration and Removal of Gills

Gills and viscera of fish harbour several spoilage bacteria in large numbers. These are potential areas where bacteria multiply and infiltrate into other areas. Partially digested food in the viscera may become sour or putrid due to bacterial action. Viscera is also the seat of very powerful digestive enzymes. They can bring about accelerated spoilage of fish. Therefore, the gills and viscera of the fish are removed before they are preserved and stored. While gills and viscera are removed, the general practice is to retain the head in tact. Evisceration should be complete with no portion of it left out. During evisceration no cut or bruise should be inflicted on the exposed belly portion. Retention of visceral parts can easily contaminate the soft belly, whereas bruises can cause accelerated spoilage by permitting easy penetration of bacteria. Removal of viscera and gills and bleeding should be done separately without contaminating other fishes. After each operation the fish should be thoroughly washed.

3.2.5 Bleeding

Bleeding is another desirable step before preserving the fish. If the fish is not thoroughly bled, blood can clot and turn dark brown affecting the colour and appearance of the meat. However, it is not possible, nor is it practicable, to bleed and eviscerate all fish, particularly the small ones. Bleeding is, therefore, restricted to only reasonably large fish like tuna, seer etc. Bleeding is done by slitting the throat of the fish followed by immersion in cold water for sufficient time, usually 30 minutes. Slitting the throat followed by hanging the fish by its tail also is practised.

3.2.6 Icing

The best method of preservation of fish onboard involves lowering its temperature. The easiest, cheapest and reasonably efficient method of lowering the temperature of fish is by icing. Ice is an effective and ideal cooling medium. It has a very large cooling capacity for a given weight or volume. It is harmless, portable and cheap. Very rapid cooling can be achieved through intimate contact between fish and small pieces of ice. About 80 per cent of fish is water and since fish freezes below –1°C, icing will maintain the fish at a temperature slightly above that at which it will begin to freeze. Ice keeps fish cool, moist and glossy, and controls deterioration.

3.2.6.1 Effect of Temperature on the Rate of Spoilage of Fish

The rate at which the fish spoils has been found to vary linearly with temperature in the range -1 to + 2.5°C. The rate of spoilage is twice as fast at 2.5°C as at 1.1°C. At 5.5°C it is twice as fast as at 0°C. Therefore, fish should not be allowed to remain exposed to sun on the deck for a long time. Exposure to sun will raise the temperature of the fish which will facilitate rapid growth of microorganisms and accelerate enzymatic and bacterial spoilage. It is also true that the iced stored shelf life of fish will be considerably reduced if the fish is kept exposed to sun for long periods in the initial stages of handling before icing. Lowering the temperature of the fish greatly retards the bacterial and enzymatic spoilage. The growth of spoilage bacteria is known to be significantly reduced by small decreases in temperature in the range -1 to + 5°C. Chilling is a very effective method of controlling spoilage in fish. The objective of chilling is to cool the fish as quickly as possible to as low a temperature as possible without freezing it. Therefore, the fish should be washed, sorted and stored without allowing any long exposure to the sun on the deck and chilled as quickly as possible. The fish shall not be trampled upon, or bruised during shovelling and handling, as this will impair the quality of the fish. There are several options for chilling fish that can be advantageously employed.

3.2.6.2 *Shelf Life of Fish in Ice*

Shelf life, in the general context, is a term used to indicate the length of time the fish will remain edible under some preserved condition, iced or otherwise. Icing does not prevent spoilage, but controls it. Fish can be stored in good condition for 3 to 15 days depending on the species and several other factors. Fatty fish like sardine has a relatively shorter shelf life of 2-5 days only in ice. Such fish will become easily rancid due to oxidation of its unsaturated fat. The belly portion also becomes very soft and gets easily ruptured due to the action of enzymes and bacteria. This phenomenon poses a serious problem particularly when the fish is caught while in active feeding stage. Belly bursting in sardine can be minimised either by dipping in brine or sprinkling solid salt over the fish before icing.

It is also known that freshwater fish have longer shelf lives than marine fish in ice; so also the tropical species keeps longer in ice than cold water species.

3.2.6.3 *Methods of Icing Fish Onboard*

Commercial fishing vessels store the catch in ice in fish holds or rooms located below the working deck. Some of the considerations in designing the fish holds are :

- should be of robust, hard and smooth surface so that it is easy to clean
- should not have cracks or crevices which will harbour dirt and bacteria
- all fitments must be strong and corrosion resistant
- the materials used for construction should not contaminate the fish
- should be adequately insulated.

Different methods are employed for storing fish in the fish holds.

• BULKING

The fish hold is divided into a number of sections or pounds using pound boards supported by upright bars or posts called stanchions. A layer of ice is spread over the bottom of the pound followed by a layer of fish and the operation is repeated until the height of fish and ice together is about 50 cm with a 5 cm layer of ice on the top. While spreading fish and ice it should be ensured that ice is spread evenly over the fish and around the edges so that the fish does not come into direct contact with the sides of the pound. Further increase in the height above 50 cm is not recommended, because with increasing height, the fish in the bottom will be subjected to excessive pressure leading to quick spoilage. Another pound board will be placed in horizontal position over the completed section which again will be supported by stanchion structure. Operation is repeated till the pound is full. Pound boards and the stanchion structures must be kept clean and out of contact with fish.

• SHELVING

The fish hold is divided into sections as in bulking, but using removable shelves, where the fish is spread in layers over ice. The lowest shelf is covered with a layer of ice about 5 cm thick. Fish is arranged in rows on the ice in thin layers and is covered with ice again. Only one layer of fish is kept on each shelf. Shelves are supported by stanchions, not by the fish and ice below. Shelving is not popular now a day because of higher space and labour requirements.

• BOXING

There are a number of designs and sizes of fish boxes made mostly of plastic and aluminium alloys in popular use. A layer of 5 cm thick ice is spread evenly on the bottom of the box followed by fish and ice in alternate layers until the box is almost full. There should be ice in the sides as well as at the top. The box should not be overfilled, nor should there be any ice protruding over the rims of the box. This is necessary to prevent crushing of the fish when the box is closed as also when the boxes are stacked one above the other in the fish hold. When boxed, the fish holds will have no shelves or their internal support.

Whether bulked, shelved or boxed, an important requirement is to ensure that sufficient ice has been placed in the bottom, sides and top. These are the areas through where more heat penetration can take place into the fish hold. Sufficient quantity of ice placed at these points can absorb the heat entering the hold. Ice should be in actual, continuous contact with all fish.

Fish are generally stored after separating them species-wise as also size-wise. However, it is also desirable that fish with different keeping qualities and shelf life under identical conditions of storage should not be mixed together.

• STORAGE IN INSULATED BOXES

Use of ice onboard vessels engaged in multi-day fishing is very common. However, many vessels, particularly those engaged in coastal fishing, are very small and do not have fish holds. It will be ideal for such vessels to carry ice in insulated boxes to preserve the catch. The size of the boxes will depend on the size of the vessel and the amount of fish usually caught in a day's fishing. The insulation required in the box will generally be governed by the ambient temperature. 10-15 cm thick expanded polystyrene insulation is in common use.

Carrying insulated boxes in daily fishing trips at least for the safe storage of low-volume high-value species like shrimp, cephalopods etc. is increasingly becoming popular. However, in cases where there is no fish hold or ice onboard and the fish has necessarily to be stored on the deck, as in the case of single day fishing with small vessels, the washed and sorted fish should be kept on the clean deck protected from sun by a covering with tarpaulin or similar material which may be kept wet by frequent sprinkling of water over it. The evaporating water from the surface of the tarpaulin will keep the fish cool for some time.

3.4 ICE NEEDS IN COOLING FISH

The quantity of ice used in preserving fish onboard should be such that it is sufficient to bring down the temperature of the fish to the desired level and can maintain it at that temperature for a sufficiently long time, at least till it is unloaded on shore. From a knowledge of the latent heat of ice, specific heat of fish, initial temperature of fish and the temperature to which it should be cooled down, the quantity of ice required can be theoretically calculated. Considering the specific heat of fish as 4J/g and latent heat of ice as 334.7 kJ/kg, the energy required to cool one kg fish say, from 30°C to 0°C will be

$$1000 \times (30\text{-}0) \times 4 = 120{,}000 \text{ J, i.e. } 120 \text{ kJ.}$$

One kg of melting ice will absorb 334.7 k J and hence the ice required will be

$$120 \div 334.7 = 0.358 \text{ kg.}$$

However, this is only a strictly theoretical calculation. This quantity is just the requirement to cool the fish to 0°C and on attaining this temperature there will be no ice left to keep the fish continuously chilled. There will be meltage of ice by the ambient air unless it is protected with an efficient insulation material. Ice is also used to chill the box or hold in which it is held. Further, the fish has to be kept cool until it reaches the consumer which needs additional inputs of ice. Therefore in the tropical climate a 1:1 ratio of ice and fish is customarily used. More ice may have to be used as required or even re-iced completely at an appropriate stage of handling.

3.4.1 Quality and Type of Ice

3.4.1.1 Quality of Ice

A potential source of contamination of fish during the primary preservation is ice itself. Irrespective of the hygienic handling and other precautions taken, the fish may get contaminated if ice of poor bacterial quality is used. The bacterial count of ice prepared out of the prescribed quality water is within 10^2 to 10^3 per g; but this may be several fold if contaminated water is used for ice making or if ice is allowed to be contaminated during its subsequent handling and storage on board, like dragging on dirty floors, or using contaminated crushers and storage boxes. Use of such ice will negate the very purpose of icing fish.

It is equally important that the ice left over after a fishing trip is not used in the next trip, even if it is unused. There are chances that the ice might have picked up bacteria during the trip with adequate chances of build up of psychrophiles.

3.4.1.2 Types of Ice

• BLOCK ICE

The conventional method of manufacturing ice is in the block form. The large block of ice is crushed to smaller pieces. Finely crushed ice is preferred for icing fish because it has a large surface area for unit mass and can cover more fish enabling its rapid cooling. However, as the crushed ice melts quickly, it will be economical to carry blocks of ice onboard and crush it only whenever needed.

In making block ice water in ice cans is placed in tanks of sodium or calcium chloride solution which is refrigerated by mechanical means. 12-24 hours may be taken to produce ice. In a modern system of rapid ice making ice blocks are produced in a few hours by improving the manufacturing process.

• RAPID BLOCK ICE

The ice blocks are formed in few hours in rapid block ice plant. In one type several tubes through which the refrigerant is passed are arranged in a tank of water. The tubes are so arranged that as ice builds up it fuses with the ice on the adjacent tubes to form a block with a number of hollow cores. The blocks are released from the tubes by a hot gas defrost.

In another type the refrigerant is circulated through a jacket around each can of water and also through pipes running through the centre of the can. Blocks are removed by gravity after a hot gas defrost.

• FLAKE ICE

Block ice, even when it is finely crushed, will have sharp contours which can cause bruises in the fish. An improvement in the ice making process which yields ice with smooth contours and in the shape of very thin flakes, usually 100 to 1000 mm^2 in area and 2-3 mm in the thickness, is the production of flake ice. Flake ice has a very high area per unit mass and can cover larger quantity of fish for a given weight compared to crushed block ice.

In making flake ice water is sprayed on to the surface of a refrigerated drum and the ice sheet formed is scraped off in the form of dry sub-cooled ice flakes. There are other models where the scraper and water spray are rotated inside a stationary drum and the ice is formed in the inner surface. The normal refrigerant temperature in a flake ice machine is -20 to -25 °C. The low temperature is necessary to produce sub-cooled ice quickly. Flake ice production is continuous, automatic and quick.

• PLATE ICE

Plate ice is made by spraying water on to the surface of vertical hollow flat plates through which the refrigerant passes. Water will freeze on the plates as flat sheet of ice. When the ice sheet attains the desired thickness it is discharged by pumping hot refrigerant gas through the plates when the ice cracks and drops away. Alternately, in some designs, water spray is used to detach ice from the plates.

Plate ice has the advantages of flake ice. The thickness of ice can be varied at will and the process is very quick.

• TUBE ICE

This is another form in which ice is produced. It is made as a hollow cylinder of about 50 × 50 mm with a wall thickness of 10 to 12 mm. Tube is formed in the inner surface of a series of hollow cylinders, the outer surface of which will be surrounded by the refrigerant. Ice released from the cylinders after defrost employing hot gas is chopped into 50 mm length. Tube ice is of a bigger size than that used in icing fish and hence will have to be crushed to the required size before icing fish. Tube ice has all the advantages of flake ice.

• SOFT ICE

Soft ice is made by freezing a weak brine or seawater in a drum provided with refrigerated walls. The crystals of fresh water ice which forms a slurry in the brine as temperature falls is pumped out into a storage tank. Ice crystals are less dense than the brine and floats on the surface and are skimmed off from there. An advantage of soft ice slurry is that it can make perfect contact with the fish. It does not cake, nor forms air pockets as can happen with other forms of ice.

3.4.2 Storage Characteristics of Fish Held in Ice

Ice cools the fish by absorbing heat from it resulting in the melting of ice. Melt water carries away a good proportion of bacteria from its surface. However, this melt water also carries away with it a considerable percentage of soluble proteins, salts and other flavouring agents as well as nutrients. The extent of spoilage of fish in ice and the accompanying chemical changes vary considerably depending upon the type and species concerned. In the case of iced stored shrimp soluble nitrogenous constituents have been found to be progressively lost from the muscle, the changes being very remarkable during the first 10 days. Glycine, the amino acid imparting sweet flavour to shrimp meat is seen rapidly lost, so also minor compounds like sugar, sugar phosphates, nucleotides etc. which are responsible for the characteristic flavour. The loss of solids by leaching depends also on the form in which fish is packed in ice, the extent of loss being significantly higher in the case of peeled shrimp and fish fillets compared to whole shrimp or fish.

In general, when stored in pens and holds with several layers of fish and ice, a weight loss averaging 7 to 8 per cent is experienced in few days. However, in control experiments where fish is stored in ice in shallow boxes, no loss but a slight gain in weight is experienced. Depth of storage affects also the external appearance adversely.

3.5 STORAGE IN REFRIGERATED FISH HOLDS

Some vessels, large in size and engaged in multi-day distant water fishing, have fish holds provided with refrigeration facility in the form of cooled grids and pipes on the roof of the fish holds in addition to insulation. It is a common practice to make use of the refrigeration facility while carrying ice to the fishing grounds to maintain the temperature of the hold below 0°C to prevent melting and consequent loss of ice.

3.5.1 'Bilgy odours'

Some fish, boxed or held in fish holds, sometimes give out some characteristic foul odour, generally called 'bilgy odours'. This is found to occur when air is completely excluded from the surface of the fish. The bilgy fish contain high percentage of anaerobic bacteria. Observance of strict hygienic conditions in the fish hold and at various stages of handling and presence of sufficient air around the fish to secure aerobic conditions can prevent the bilgy odours. Packing fish in boxes or holds with crushed ice in the form of small lumps can leave sufficient air pockets within the pack and can be considered as a measure to be adopted to control bilgy odours.

3.6 ALTERNATIVE CHILLING METHODS

Ice is a versatile cooling medium for fish. However, under certain circumstances it will not be practicable to ice the whole catch. An example is the heavy catch of pelagic shoaling fish over a short time in purse seines. Carrying huge quantities of ice for efficient icing of such fish is often not practicable. Handling such large quantities of fish for icing and storing will consume a lot of crew time which can otherwise be spent on fishing efforts. In such cases it is an ideal method to stow the fish in seawater cooled to temperature near to that of melting ice.

3.6.1 Preservation in Refrigerated Seawater (RSW)

Seawater contains 3-4 per cent sodium chloride and freezes at about -2°C. By refrigerating seawater it is possible to maintain a temperature just above its freezing point. Fish stored in this water can be maintained at -1°C to 2°C throughout the storage period. Seawater cooled by mechanical means (refrigerated seawater, RSW) is circulated through tanks installed onboard the fishing vessels. Fish is kept in the tanks, and being less dense than the seawater, floats in it. Fish can also be suspended in perforated vessels immersed in RSW.

A schematic representation of RSW system is shown in Fig. 3.1

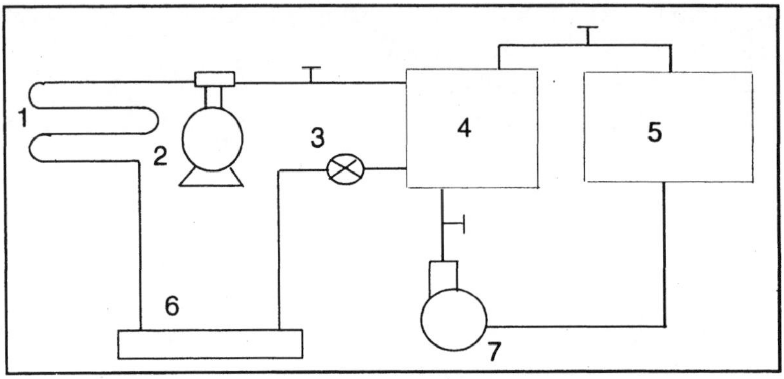

Fig. 3.1 : RSW system (1) condenser, (2) compressor, (3) expansion valve, (4) chiller, (5) storage tank, (6) receiver, (7) circulating pump

Chilling fish in RSW is faster than chilling in melting ice because of the more intimate contact between the fish and the cooling medium. This method of preservation has been found quite satisfactory for non-fatty as well as big size fishes. Such fishes can be stored in RSW for longer periods than in ice. Another advantage is that more fish can be packed per unit volume compared to preservation by icing causing no undesirable physical damage to fish.

However, with fatty fish like oil sardine, the rapid development of rancidity in RSW is a limiting factor. One method of overcoming this disadvantage is by bubbling carbon dioxide gas through the water. Meat of some fish like mackerel may become blackened on storage in RSW. Uptake of salt by fish is considered another disadvantage as it often reaches objectionable levels. The salt uptake, however, will depend on the size and species of fish, whether the fish is gutted or whole, ratio of fish to water and the length of storage. Further, the offal from such fish will be unsuitable for fishmeal production. The eyes of fish stored in RSW become opaque and the gills become bleached earlier compared to iced fish. However, they have greater firmness and better appearance than iced fish. RSW storage is not a method recommendable for preservation of shrimp as it develops black spots. But for this, shrimp keep well in RSW.

Some bacteriological problems are also associated with RSW storage. Even at the low temperature of operation the liquid in the tank acts as a favourable substrate for bacterial build-up.

With an efficient refrigeration system there is a possibility that the fish may be cooled to or below the temperature at which they freeze. This should be avoided by controlling the temperature of water. Freezing is also likely to render the eating quality of fish poor. However, partial chilling can prevent penetration of salt into the meat.

3.6.2 Preservation in Chilled Seawater (CSW)

Another method of storing fish onboard is in seawater or dilute brine made upto the concentration of seawater by mixing it with ice, generally referred to as chilled seawater (CSW). This is a very simple method. There are also instances where it is advantageous to combine RSW and CSW systems where some cooling is provided by refrigeration and some by ice.

In RSW system it is necessary to circulate the water by pumping to ensure even mixing and cooling. This may sometimes be done with CSW systems also, but pumping is generally not necessary, as the motion of the vessel itself will provide adequate mixing. Another advantage of CSW is that freezing fish, which is a possibility in RSW, does not occur here. But for these, the advantages and disadvantages of chilling by RSW and CSW are more or less same.

3.6.2.1 Quantity of Ice Required to Chill Water

In RSW or CSW storage the recommended ratio of fish to water is between 3 : 1 and 4 : 1. The quantity of ice required to provide the necessary extent of cooling in a CSW system can be calculated as follows.

Consider 4 tonnes (4000 kg) fish at 30°C is to be cooled to -1°C where the fish to water ratio is 4 : 1. Fish and water together will weigh 5000 kg. The amount of heat to be removed is

> 5000 × (30 –(–1) × 4J (4J as specific heat of both fish and ice) = 620,000 kJ

One kg of melting ice will absorb 334.7 kJ. Therefore, the weight of ice required is

620,000 ÷ 334.7 = 1852.4 kg.

Thus about two tonnes of ice will be required to provide the initial cooling to reduce the temperature of fish and water to -1°C. Additional ice will be needed to compensate for the heat leak. The quantity of ice needed for this will depend on the adequacy of insulation of the tank, the ambient temperature, duration of storage etc. A calculation of this sort is not so easy with respect to RSW systems.

3.7 FREEZING PRESERVATION OF FISH ONBOARD VESSEL

Fish has a limited shelf life of 3 to 10 days in ice or in RSW/ CSW storage depending upon the species and nature of fish. If a longer shelf life exceeding several days is desired, freezing is, perhaps, the best alternative, particularly in cases where the fish is intended for canning. Tuna, salmon etc. are examples of fish preserved onboard by freezing for canning later.

3.7.1 Brine Freezing

Brine freezing systems are largely employed for freezing fish onboard. Saturated brine is used in such systems as it can attain a temperature as low as -21°C before it starts freezing. Refrigerated brine is circulated through tanks in which the fish is held. Time taken to freeze the fish will depend on the temperature of the brine and size of the fish. Under ideal conditions it may take 15-20 minutes. Once frozen, the fish is removed from the brine tank and is stored in cold rooms.

The main advantage of brine freezing is the rapidity with which heat is removed from the fish because of the intimate contact of the fish with brine from all sides. Further, salt solution has an inhibitory effect on the microorganisms when salt is present in hypertonic concentration. One probable disadvantage of the system is the likely absorption of salt by the fish when held for longer periods in contact with brine at temperatures above the freezing point of brine. However, this is not a serious disadvantage for fish intended for canning as at some stage or other in the canning process the fish is salted. Excessive salt penetration can be controlled by using a solution of a mixture of glucose or corn syrup and salt as freezing medium.

3.7.2 Vertical Plate Freezer

Apart from brine freezing other methods employed for freezing fish onboard include the use of vertical plate freezer or blast freezer. Vertical plate freezer consists of a series of vertical refrigerated plates with space between them called stations. The refrigerant passes through the plates and reduces their surface temperature to -30 to -40°C. The fish are often chilled before loading into the freezer. Fish is dropped into the space between the plates until each station is filled. The plates are then closed together to make fish into blocks. Fish of similar size should be packed in each station; dissimilar fish will lead to uneven freezing. After freezing the blocks are released from the stations by a partial defrost and are held in the fish holds.

Advantage of vertical plate freezer is that no freezing trays are required; operation is simple and for a given output it requires less space and is easy to install and operate.

3.7.3 Air Blast Freezing

This type of freezers consist of an insulated room where the fish is kept and refrigerated air is blown over it with the help of a blower. Cold air while passing over the fish picks up heat from the fish and freezes it. Fish of any shape and size can be frozen, but occupies more space and consume more energy which limits its use onboard.

3.8 HYGIENE IN FISH HANDLING

The major cause of spoilage of fish is by the action of bacteria. If the fish is contaminated with pathogenic bacteria it can cause various illness in the consumer, some of which may result in death. The best way to minimise bacterial contamination and their further proliferation in fish is to ensure hygienic handling and keeping the fish clean and chilled. Some of the important requirements in hygienic handling of fish are detailed below.

3.8.1 Supply of Clean Water

Adequate supply of clean water is one of the most important requirements in handling fish. Washing with water removes more than 90 per cent of the surface bacteria. However, the water used should have been filtered and chlorinated to ensure safe bacterial level; otherwise, it may by itself turn out as a source of bacterial contamination including contamination by pathogenic bacteria. This will create a situation where washing will do more harm than good. Water used should preferably be of potable grade having residual chlorine not more than 5 ppm. Water for cleaning of premises, utensils etc. should have a high level of residual chlorine, 10-100 ppm.

3.8.2 Personal Hygiene

Hygienic handling of fish onboard should start with strict observance of hygiene of the persons involved. Fish can pick up bacteria from several sources during handling including the persons handling them. It is, therefore, very important that the persons handling fish should maintain strict hygiene and be free from any communicable disease. While handling fish they should wear clean uniform including headgear, and also should use waterproof coats or aprons and gumboots. Before start of handling, the persons involved should wash their hands from elbow downwards using soap and water followed by disinfection in adequately chlorinated water. This should be repeated after each break of work, visits to toilets etc. or whenever there is a chance to contaminate the hands. Those having bruises on palm, fingers etc. should not handle fish because these are sources of bacteria whose presence in food is considered undesirable from the hygienic point of view. They should also keep their fingernails neatly trimmed. Spitting, chewing tobacco, smoking and the like should be prevented in the areas of fish handling.

3.8.3 Cleaning and Disinfection

Thorough cleaning and disinfection of the boxes, all handling equipment, utensils etc. are primary requirements in hygienic handling of fish. All these shall be made slime-free and clean using appropriate detergents and disinfectants. This should be done at the end of each working day or work schedule. Use of dock water for washing is dangerous, as it will generally be grossly infected. Chlorine is the generally used disinfectant. For metallic surfaces water containing 100 ppm available chlorine will be sufficient, but for wooden surfaces it can go up to 1000 ppm. After a contact period of 15 minutes the disinfectant may be washed off using clean water.

Cleaning can also be done by hosing water under pressure. Clean seawater containing the requisite level of chlorine will be more effective than plain untreated water in removing slime from the deck. Such washing programme may be resorted to at the start of the fishing trip as also after unloading the catch on the deck and its clearance therefrom. Improper cleaning will allow build up of slime and bacteria, which in turn will contaminate the fish coming in contact with it.

Suggested Reading

Bramsnaes, F. (1965) in Borgstrom, G. (ed.) *Fish as Food* Vol. IV. Academic Press, New York and London.

Burgess, G.H., Cutting, C.L., Lovern, J.A. and Waterman, J.J. (eds) (1967) *Fish Handling and Processing.* Chemical Publishing Co. Inc. New York.

Cutting, C.L. (1953) *Fish Saving : History of Fish Processing from Ancient to Modern Times.* Leonard Hill Books Ltd., UK.

FAO (1975) *Ice in Fisheries* . FAO Fisheries Report No. 59. FAO, Rome.

FAO (1981) *Refrigerated Storage in Fisheries.* FAO Fisheries Technical Paper No. 214, FAO, Rome.

IIR. (1976) *Cooling and Freezing Aboard Fishing Vessels.* International Institute of Refrigeration, Paris.

Lima Dos Santos, C.A.M. (1981) The Storage of Tropical Fish in Ice: A Review. *Tropical Science,* 23 : 97-127.

Myers, M. (1981) *Planning and Engineering Data. 1. Fresh Fish Handling.* FAO, Fisheries Circular No. 735, FAO, Rome.

Waterman, J.J. (1964) *Bulking, Shelving or Boxing?* Torry Advisory Note No. 15. Central Science Laboratory, Aberdeen.

4
DRYING AND DEHYDRATION

DRYING is one of the oldest known methods of preservation of food. Though the technology of food preservation and processing has undergone revolutionary changes over the years and several new products processed employing diverse techniques have made their firm presence in the market, drying still continues to be the most widely used method for preservation of several foods including fish. It is also considered the least expensive method of food preservation.

4.1 DRYING *VS* DEHYDRATION

4.1.1 Sun Drying

Drying involves removal of water from a body, in our context fish. Traditionally, fish used to be dried under sun and the term 'drying' has come to imply drying under the sun. Sun drying is carried out in the open air using the solar energy to evaporate the water in the food. The evaporated water is carried away by the natural air currents. The efficiency of the process and the quality of the product remain at the mercy of the elements of nature and, therefore, the product suffers from serious disadvantages, some of which are:

- dependence on weather; the operations can be carried out only when bright sun light is available
- long duration of drying; under unfavourable conditions of weather drying may take several days to complete
- there is no control over the operating parameters

- possibility of contamination with dust, sand etc.
- possibility of infestation with insects, their eggs and larvae
- poor quality of the product
- short shelf life.

However, it must be emphasised that the process of drying employed might have met the needs of the people of the time for preserved foods to sustain them during periods of deprivation.

4.1.2 Dehydration

One of the aims of preservation and processing a perishable food is to avoid the wastage at the time of plenty and to make it available at a later time when its availability is limited or it is totally non-available. However, while achieving preservation care has to be taken to ensure that the wholesomeness, taste, texture, physical appearance etc. are also maintained to the maximum possible extent. Over the years several improvements have been brought about in the process of drying keeping in view the above consumer requirements. An important achievement in drying in this direction is the development of artificial driers where the important operational parameters like temperature, relative humidity and velocity of air etc. can be controlled. This has led to 'dehydration' which refers to a process of drying under controlled operational parameters like temperature, air velocity and relative humidity. In order that these parameters are controlled, the drying has necessarily to be carried out in an enclosed atmosphere. Therefore, in 'dehydration' the process is carried out in an enclosure provided with facilities to control the operational parameters. Most of the disadvantages encountered in sun drying can, more or less, be overcome and a product of desired quality and reasonable shelf life can be obtained by dehydration. However, the terms 'drying' and 'dehydration' are being used now a days without much specificity.

4.1.3 Advantages of Dried Foods

- dried and dehydrated foods are highly concentrated foods compared to any other preserved form of foods
- drying reduces the microbial activity and thus reduces the spoilage of foods due to microbial activity
- with reduced water content enzymic and many chemical processes are retarded
- dried foods are less expensive to produce
- there is no involvement of complicated machinery and equipment for processing and packaging
- they are stable at most ambient temperatures
- distribution costs are minimum

4.2 PRINCIPLES OF DRYING

Water is essential for the activity of all living organisms including microbes. Reduction in the water content or its complete removal will proportionately retard or totally stop all microbial and autolytic activities, thus preventing spoilage and resulting in preservation.

4.2.1 Water Content and Related Activity of Micro-organisms

Water content of fresh non-fatty fish is around 80 percent; in fatty fish this may be less by a quantity equivalent to the percentage of fat. When the water content is reduced to a level below 25 percent bacterial action stops. Further reduction to below 15 percent can prevent mould growth. However, it is a common experience that dried fish containing relatively less water will spoil more quickly due to the action of bacteria than a similar sample containing more water when the latter has been salted before drying. Therefore, the expression 'water content' alone cannot satisfactorily explain the survival and growth of microorganisms in dried products. The relatively less growth of microorganisms in a substrate with a higher water content depends upon its composition.

4.2.2 Water Content vs Water Activity (a_w)

The bonds between water and protein molecules in fish may be in several forms. Three layers of water, an adsorption layer, a diffusion layer and a free layer, surround the surface of a colloidal particle. Water at the adsorption layer that is tightly bound to the colloidal particle is called 'bound water'. The diffusion layer is less tightly bound to the adsorption layer. The farther it is from the adsorption layer, the more it behaves like free water. The third layer, the 'free water' has all the properties of ordinary water and can enter into all normal chemical reactions and support microbial spoilage. It is this 'free water', i.e. the water available to support microbial growth and react chemically, that is important in the drying process. The most effective physical method of assessing the state of water in foods is the measurement of partial vapour pressure of water in equilibrium with a given moisture content at constant temperature. This is called water activity, a_w. The a_w is a measure of water available to support microbial growth and chemical reactions. It is an important determinant of microbial growth and metabolic activity as well as of food deteriorating enzymes and chemical activity. Majority of food preservation procedures employs modification in the state of water in the foods. Water activity is expressed as the ratio between the water vapour pressure on the surface of the product and the water vapour pressure of distilled water, both at the same temperature.

$$a_w = \frac{\text{water vapour pressure on the surface of the product}}{\text{water vapour pressure of distilled water}}$$

Water activity of pure or distilled water is assigned the value 1.0 and of an absolutely dry product 'zero'. Water activity in a food is expressed as a fraction relative to pure water. The a_w of salted fish will be low because of the presence of dissolved salt and this will explain why salted fish with relatively high moisture content will be more stable whereas unsalted fish with lower moisture content will spoil faster.

4.2.3 Water Activity and Related Microbial Growth

Fresh fish has a water activity above 0.95 and as it is lowered, microbial activity becomes gradually inhibited. Limiting a_w for the growth of various microorganisms are presented in Table 4.1.

Table 4.1 : Water activity (a_w) and related growth of microorganisms

a_w	Microorganisms inhibited
1	- none
0.95	- Gram-negative rods
0.91	- most spoilage bacteria - cocci, lactobacilli etc.
0.88	- most yeasts
0.80	- most moulds, *Staphylococcus aureus*
0.75	- most halophilic bacteria
0.65	- xerophilic moulds
0.60	- osmophilic yeasts

4.2.4 Measurement of a_w

One of the simplest methods of measurement of a_w of a food is in terms of its vapour pressure manifested as % RH generated in equilibrium with the food in a closed system at constant temperature. a_w is equal to the equilibrium RH divided by 100. A sample of food of known moisture content is allowed to equilibrate with a small headspace in a tight enclosure and then the equilibrium relative humidity (erh) is measured. The a_w of food = erh (air) ÷100.

4.3 DRYING PROCESS

Drying process involves essentially two steps: migration of water from the wet parts of the material to a vapourisation zone where the water vapourises and its removal from there to the surrounding atmosphere by the circulating air. These processes are called diffusion and evaporation respectively. The efficiency of a dehydration system will depend on how these factors are controlled.

4.3.1 Factors Influencing Dehydration

4.3.1.1 Equilibrium Moisture Content and the Influence of Humidity

The pressure of water vapour at the surface of the fish is different from that of water in the air. Because of this difference in partial pressures migration of water molecules takes place from one medium to the other. The partial pressure of water vapour at the surface of fresh fish, which is saturated with water, is greater than that of the water vapour in the air. Therefore, migration of water vapour from fish to the surrounding air, i.e. dehydration, takes place. The converse will be true if partial pressure of water vapour at the fish surface happens to be less. The processes of sorption (uptake of water) or desorption (release of water resulting in dehydration) will continue until the pressure of water vapour at the fish surface and in the air is equalised. The moisture content in the fish is then in equilibrium with the air humidity. The moisture content, which is in equilibrium with the atmospheric humidity, is called the equilibrium moisture content.

Equilibrium moisture content is a very significant factor in a drying operation, particularly in a controlled operation. Equilibrium moisture content of fish under drying will vary depending on the variations in the vapour content in the air. Any method of drying by exposure to air can remove only that moisture in the fish, which is over and above the equilibrium moisture content possible at the operating air temperature and humidity. Under a given set of conditions the drying rate can be increased only by effecting a shift in the equilibrium moisture content by dehumidifying the air in the drying atmosphere.

Salting lowers the partial pressure of water at its surface, therefore its equilibrium moisture content will be high. The saltier the fish is, correspondingly less will be the partial pressure of water at the surface and higher the equilibrium moisture content.

4.3.1.2 *Heat Transfer*

Drying of fish involves the removal of water from the solid fish body with the help of an input of thermal energy. The outward flow of water vapour is linked with an inward flow of heat. Thus drying is linked with the rate of supply of heat to the point from where evaporation occurs and its conduction within the material. Transfer of heat by conduction takes place by the exchange of momentum of random thermal motion of the molecules to the neighbouring molecules, i.e. passed along hand to hand. However, the heat transfer should also consider provisions of latent heat of evaporation.

4.3.1.3 *Mass transfer*

As has been mentioned earlier drying is a phenomenon involving migration of water within the body to the surface and conveyance of vapourised water away from the surface. The factors controlling the rate of transfer of water from the moist body to its surroundings are the physical characteristics of the body's environment, especially the temperature, pressure, humidity and air velocity. The factors, which determine the rate of movement of water within the body, are mostly independent of the external conditions.

Some of the physical mechanisms which are considered as involved in mass transfer are :

- liquid movement under capillary force
- diffusion of water caused by a difference in concentration
- surface diffusion in liquid layers adsorbed at solid interfaces
- water vapour diffusion in air filled pores caused by a difference in partial pressure
- water vapour flow under differences in total pressure as, for example, in the vacuum drying under radiation
- water flow caused by shrinkage and pressure

gradients
- flow caused by gravity
- flow caused by vaporisation-condensation sequences.

4.4 FACTORS INFLUENCING DRYING RATE OF FISH

Rate of drying is influenced by several factors-the intrinsic nature of the fish being dried and several external factors employed in the drying atmosphere.

4.4.1 Nature of the Material Being Dried

The individual nature of the fish, its chemical composition and physical structure are important factors influencing the rate of drying. Under constant external conditions higher rate of drying will be exhibited by the fish

- containing more water
- having larger surface area in relation to weight. Splitting the fish will result in increase of surface area relative to weight and the rate of drying will become faster.

However, the presence of high mounts of fat will retard the rate of escape of water and hence, the rate of drying.

4.4.2 Relative Humidity (RH) of the Air

Absolute humidity (H) is the mass of water contained by a unit mass of air.

$$H = \frac{\text{water vapour (kg)}}{\text{dry air (kg)}}$$

Relative humidity (RH) is the absolute humidity divided by the humidity of saturated air under identical conditions of temperature and pressure assuming unit mass of dry air in both cases. In other words, relative humidity refers to the degree of saturation of vapour space or body of moist air and is expressed in per cent saturation of air with moisture at a given temperature. The most important single factor correlated with drying rate is the relative humidity of the drying air passing over the fish under drying. RH is measured in terms of wet bulb depression. Wet bulb depression is the difference between the temperature readings of two thermometers kept in an air current, one with its bulb kept damp (wet bulb) using a wetted muslin sleeve and the other uncovered and dry (dry bulb). In the initial stages of drying this corresponds to the difference between the temperature of wet fish surface (wet bulb) and that of the drying air. The difference between dry bulb and wet bulb temperatures is related to the RH. When the difference becomes zero it implies that the air is saturated with water vapour (RH 100%) and it cannot carry any more water and hence the fish will not dry any further. As drying progresses in a closed system eventually such a situation arises. A change in air temperature is usually accompanied by a change in RH. By increasing the air temperature its moisture carrying capacity can be increased; however, this may adversely affect the quality of the product.

During the later stages of drying, the rate of internal diffusion of moisture becomes less than the rate at which it is evaporated from the surface. At this stage it is desirable to maintain the RH of the air at a relatively higher level in order to avoid over-drying of the fish surface and thus to avoid or minimise the occurrence of case hardening.

4.4.3 Air Temperature

In the beginning of the drying cycle evaporation of water produces a cooling effect on fish and air around and hence the temperature of the fish tends to fall below the ambient. Difference in temperature does not show any appreciable influence on the drying rate at this stage. Neither the wet bulb depression nor air velocity significantly affects the drying rate. To maintain constant temperature of drying the heat used for evaporation has to be fed back to the system. For every kg of water to be evaporated 550 kcal (2600 kJ) of heat must be supplied to the system.

The influence of temperature on drying rate becomes more pronounced in the later stages of drying when the moisture content falls. In the low moisture range the drying rate is so slow that the cooling effect of evaporation is not appreciable. The fish being dried attains very nearly the temperature of the drying air. The internal redistribution of water is the true factor which determines the drying rate in this phase. This can be accelerated by a rise in the material temperature.

The recommended temperatures for drying fish of temperate waters is 25 to 35°C whereas tropical fish is generally dried at 40-50°C. Higher temperature may cause cooking of fish flesh which will make the resultant dried product brittle. Therefore it is important that the temperature employed at any stage of drying is not high enough to result in cooking of fish flesh.

4.4.4 Air Velocity

The rate of drying will be faster with higher velocity of air passing over the fish. In a drying system three layers of air can be considered to be prevailing over the fish - a stationary layer close to it, a slowly moving layer outside this and a third outer turbulent layer of air. The stationary layer next to the fish becomes saturated with moisture, which passes into the slowly moving layer. When the air velocity of the outer layer is high, the slow moving air will become thinner allowing more rapid movement of water away from the fish surface.

The air velocity should be maintained such that the processes of diffusion and evaporation proceed smoothly. Air velocity above a certain range does not have any beneficial effect on the rate of drying. In the later stages of drying when rate of diffusion of moisture becomes less than the rate at which it evaporates from the surface, a lower air velocity at a level which permits adequate circulation and heat transfer will be desirable.

At an air velocity of 70 m/min the drying rate is twice as rapid as in still air. It is three times as rapid as in still air when the air velocity employed is 140 m/min. It is considered best to employ an air velocity between 75 and 130 m/min for fish drying.

4.4.5 Constant and Falling Rate Drying

Non-hygroscopic solids, in general, exhibit two distinct phases of drying; the constant rate drying and the falling rate drying.

4.4.5.1 Constant Rate Drying Period

In the beginning of drying the fish surface is saturated with water and evaporation takes place as from a free water surface. The rate of escape of moisture per unit weight per unit time will remain a constant as long as the surface remains wet. This is governed by conditions of the surrounding air like temperature, RH and velocity which will influence how rapidly the air can supply heat to the water in the fish and remove the vapour produced. During the pendency of this period heat absorbed for evaporation of water at the moist surface is balanced by heat flow from the warm air stream into the cool wet body and both are in equilibrium. In physical terms water is diffusing to the surface of the fish at a rate equal to its removal from the surface. The temperature of the fish being dried, generally, is the wet bulb temperature of the air in contact with it.

4.4.5.2 Falling Rate Drying Period

At a stage in the drying process, water can no longer diffuse to the surface, as rapidly as it is evaporated and, therefore, the surface no longer remains wet. The moisture content at this stage called the "critical moisture content". The rate of drying then will depend on the rate at which water in the fish migrates to its surface. The zone of evaporation recedes deeper and the rate of movement of moisture continues to decrease; the drying rate falls progressively. The rate of drying is controlled by the rate of diffusion of moisture which is influenced by several factors

- composition of fish - high fat content retards the rate of drying.
- thickness of fish - with thicker fish water in the middle layers has to travel long to reach the surface.
- temperature of the fish - diffusion of moisture from the deeper layers to the surface will be greater at higher temperatures. This is the only aspect which can be controlled to derive beneficial effects during the falling rate period.
- moisture content - when the moisture content becomes low the rate of movement to the surface is reduced.
- amount of added salt - the higher the amount of added salt, the more slowly the moisture diffuses to the surface.

However, the drying rate at this phase does not depend on the humidity of the air, provided the air is not saturated. The drying rate also does not depend on the speed of the air passing over the fish. This phase of drying is called the falling rate period. The solid materials in the fish begin to absorb heat from the air and the temperature of the fish begins to approach the dry bulb temperature of the air. A typical drying rate curve of fish is shown in Fig. 4.1.

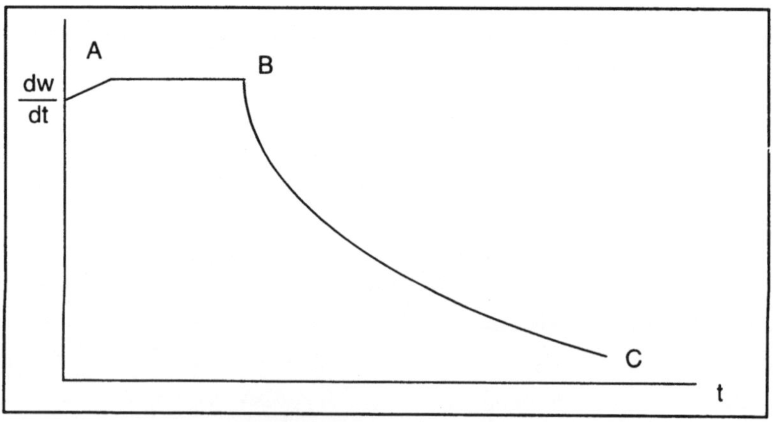

Fig. 4.1 : Drying rate curve
dw = difference in weight, dt = difference in time, t = time, AB = constant rate period, BC = falling rate period

In many cases the fish is dried after salting when a substantial quantity of body water is lost through osmosis. Salted fish, therefore, does not exhibit a constant rate drying period. A continuously falling drying rate is exhibited by such fish during dehydration.

4.4.6 Effect of RH of Air on the Final Water Content of Fish

During drying under a given set of operational conditions the loss of moisture stops at some point and it becomes impossible to remove any further water from fish. Relative humidity of the air in the drying atmosphere is the factor controlling this phenomenon. The minimum moisture content obtainable for lean fish at different RH levels is given in Table 4.2.

Table 4.2 : Relation between RH and minimum moisture content obtainable in lean fish

RH of the air %	Minimum moisture content obtainable in fish (%)
20	7
30	8
40	10
50	12
60	15
70	18
80	24

4.5 FISH DRYING PROCESS

4.5.1 Pre-process Operations

Before drying, the fish is often subjected to different pre-process operations. These are mainly dependent on the size and nature of the fish, as also the end product desired.

4.5.1.1 Cutting/Splitting

Conventionally, very small and thin fishes like anchoviella are dried whole without salting. Small fishes like sole, small croakers, anchoviella etc. are salted whole and dried. However, gutting and cleaning before salting will reduce spoilage and improve the quality. Splitting open or cutting into pieces before salting is necessary in the case of bigger fishes. In the case of very big cat fish, shark, rays etc. they are split open and deep scores are made in the exposed flesh before salting. This will increase the area of exposed flesh for greater contact with salt, permit easy penetration of salt and increase the surface area for evaporation of moisture, thus bringing about a reduction in the time required for drying.

4.5.1.2 Salting

Fish may be dried without salting or after salting; it is better that the fish is immersed in brine rather than in dry salt so that the required concentration of salt in fish flesh is achieved in a shorter time. When fatty fish is salted, air should be excluded during brining to reduce the rancidity. It is also desirable to employ saturated brine to enable easier uptake of salt by the fish flesh.

4.5.2 Natural Drying

Solar and wind energies are made use of in natural drying process. Natural drying has been known to yield very stable dry fish and several such products are known to have remained safe for years together. Some of the essential requirements for production of high quality dried fish by natural drying in the tropics are:

- sufficiently high air temperature. Temperature in the range 35-40°C will be ideal. In many tropical countries temperature often becomes higher.
- sufficiently low RH to permit drying fish to an a_w level not conducive for bacterial spoilage. RH above 70-75% will not help dry the fish to the desired level. Salted fish will tend to absorb moisture from the surrounding air if the RH rises above 75%. In coastal regions the humidity will be often very high limiting the speed of drying.

- use of raised platforms. Air movement at ground levels is comparatively slow. Better air movement can be ensured if the fish is kept raised by about one metre above the ground.
- use of drying racks. Keeping the fish on racks kept above ground level will facilitate movement of air both under and over the fish, thus allowing drying from both upper and lower surfaces. Contamination of fish by dust or sand also will be minimised. Racks with sloping tops will allow for easy draining of any surplus water on fish surfaces in the beginning of drying.

Diagrams of some drying racks are showin in Fig. 4.2

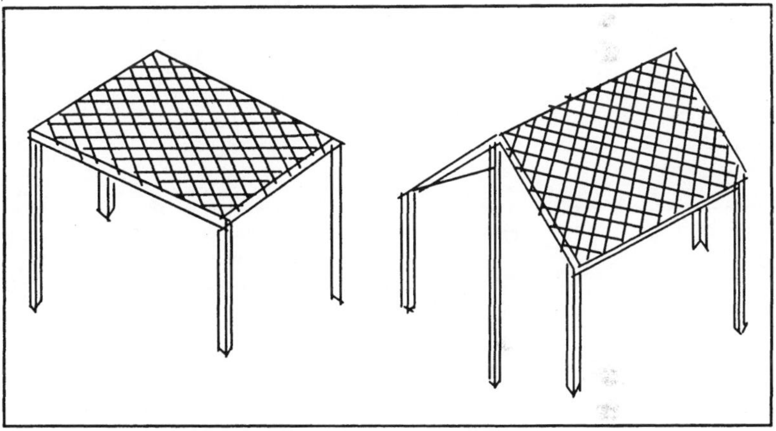

Fig. 4.2 : Drying racks

4.5.3 Mechanical Drying

Mechanical driers can be broadly classified into two types. In one type, the heat is transferred into the product through a hot gas, usually air. A heat/water vapour exchange takes place at the contact point of hot air with the product and the outgoing air will carry the water vapour away. Kiln drier, tunnel drier etc. are examples of this class of driers.

In the second type, heat is transferred to the product through a solid surface which may also be used as the carrier for the product to be dried. A typical case is a vacuum shelf drier where the hollow shelves serve both as heat transfer medium and also hold the material to be dried. The whole drying cabinet is evacuated and the water vapour produced is removed using a vacuum pump. However, in all such types of drying the product need not necessarily be held under vacuum, it can be exposed to air as well.

4.5.3.1 Cabinet Drier

This is a simple batch operation model drier used for relatively small scale operations. A typical cabinet drier may consist of an insulated or non-insulated framed structure. Material to be dried, uniformly spread in trays may be placed on permanent supports provided in the drier. A fan located inside the drier will blow air past a heat source which pass across or through the material loaded in trays (Fig. 4.3)

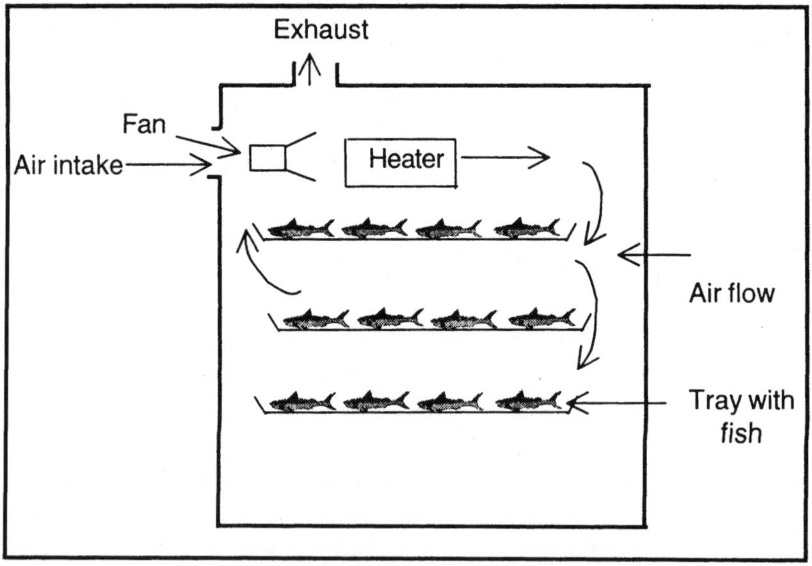

Fig. 4.3 : Cabinet drier

4.5.3.2 *Kiln Drier*

Kiln drier is a batch drier. A typical drying kiln will consist of a two-story building. The floor of the upper story is slotted or may be composed of narrow slats on which the material can be spread. This story serves as drying room. The burners or furnace producing hot gas is located in the lower floor. The hot gas passes through the product by natural conduction; often forced circulation with the help of a fan also may become necessary. The material being dried has to be turned and stirred frequently to ensure uniform drying (Fig. 4.4).

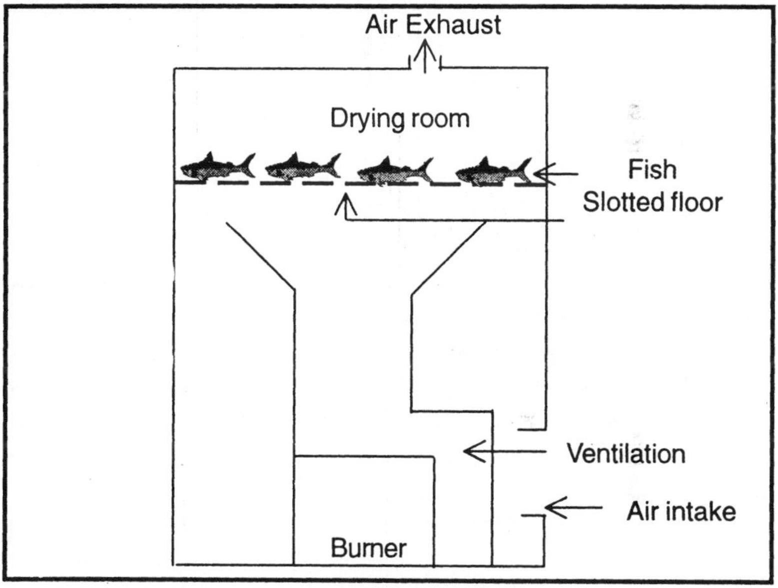

Fig. 4.4 : Kiln drier

4.5.3.3 *Tunnel Drier*

Tunnel driers are most commonly used for drying fish. These are made in the form of long tunnels, 10-15 m long. Trolleys loaded with trays containing fish are moved at a predetermined schedule through the tunnel. Hot air is blown over the material across the trays. For drying fish using the tunnel driers the production cycles are so planned that when a trolley of fresh fish leaves the drier at one end, a fresh trolley of fish to be dried is introduced at the other end.

• PARALLEL FLOW DRYING TUNNEL

In tunnel drying the movement of air can be maintained in either direction in relation to the flow of the material. In some systems the air movement is in the same direction as that of the product movement. This type of drier is called a parallel flow drying tunnel. In such driers the hottest air comes in contact with the wettest fish. Therefore relatively higher temperature can be employed for drying. One serious disadvantage of the system is that the air towards the outlet side will become cool and highly humid; therefore the finished product may not be sufficiently dried.

• COUNTERFLOW DRYING TUNNEL

The movement of air in this type drier is in the direction opposite to the movement of the product. The hot dry air comes first in contact with the driest material so that the finished product obtainable will be very dry. However, in prolonged drying schedules the fish from the other end of the tunnel may remain in humid warm air for long periods without getting sufficiently dried. Bacterial and other types of spoilage are likely to take place in the fish during this period. Dried product also should not be left to remain in the drier for longer periods than necessary. Otherwise, the product may become too dry and will affect the organoleptic characteristics.

Fig. 4.5 and 4.6 show the schematic presentations of parallel and counterflow drying tunnels.

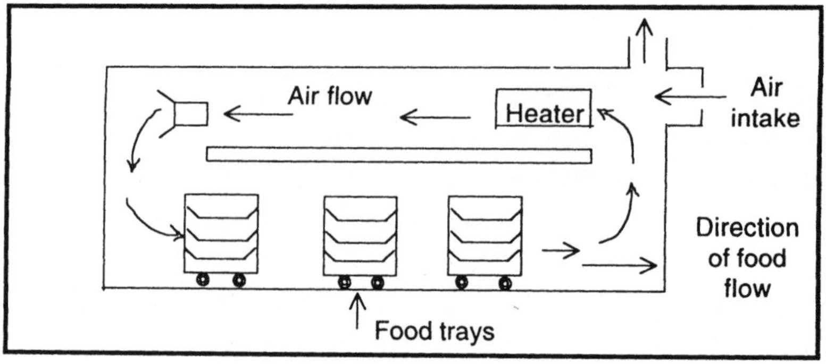

Fig. 4.5 : Parallel flow tunnel drier

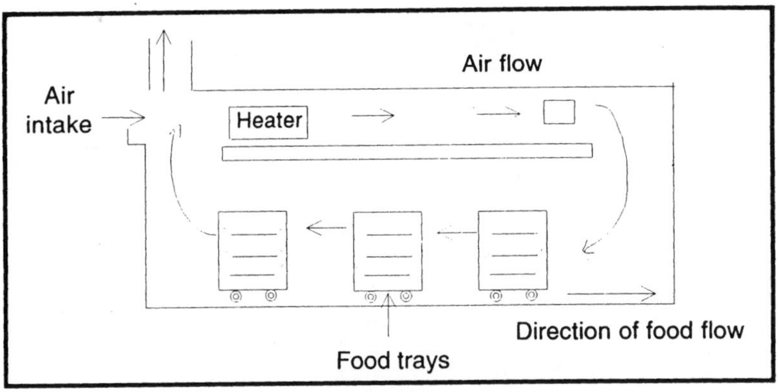

Fig. 4.6 : Counter flow tunnel drier

Tunnel driers may be designed based on the concept of hot air recirculation or allowing the hot air passing over the fish to escape to the atmosphere. The recirculating air will become highly humid to the extent of slowing down the drying significantly. Therefore, it is very essential that tunnel driers with provisions for recirculating hot air should have adequate facilities for dehumidification and control of humidity in addition to control of air temperature.

4.5.3.4 *Spray Drier*

Spray driers are generally used for drying foods which are in the form of liquids or suspensions. In principle a food in a liquid or paste form is atomised and dispersed as minute droplets which are suspended in a stream of hot air in a chamber where it gets rapidly dried. The dry particles suspended in the air stream flow into separation equipment where they are separated from the air, collected and packaged. In the application of spray drying to fish products it is limited to products like fish protein hydrolysates and fish powders.

Fig. 4.7 shows the features of a simple spray drier.

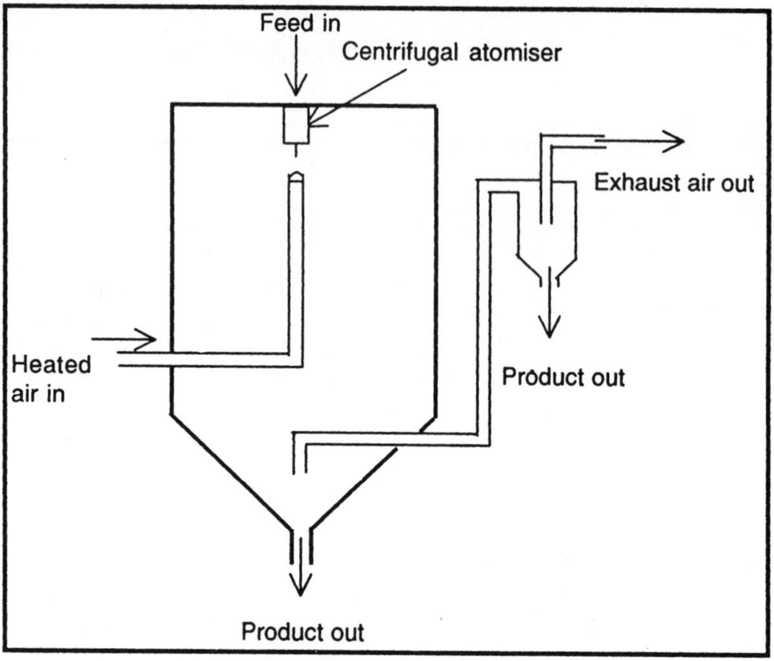

Fig. 4.7 : Spray drier

Better retention of colour, flavour and nutritive value are the advantages associated with spray dried foods.

4.5.4 Heat Transfer Through Solid Surface

4.5.4.1 Drum Drier

Drum drier is an example where heat transfer to the product takes place through a solid surface. Drum drier is used for drying fluid materials. The food product in the form of slurry is deposited as a thin film on the drum. The drum is heated, generally by steam; while it is being rotated. Drying can be done keeping the drum open to the atmosphere. If the material is desired to be dried under vacuum, the drier can be enclosed in a vacuumed chamber. The product when dry is removed from the surface of the drum using a scraper blade.

Drum driers are classified as single drum, double drum and twin drum types. Single drum drier comprises only one roll. Double drum comprises two drums rotating towards each other. Twin drum is similar to double drum, but rotate away from each other. Different types of drum driers are illustrated in Fig. 4.8.

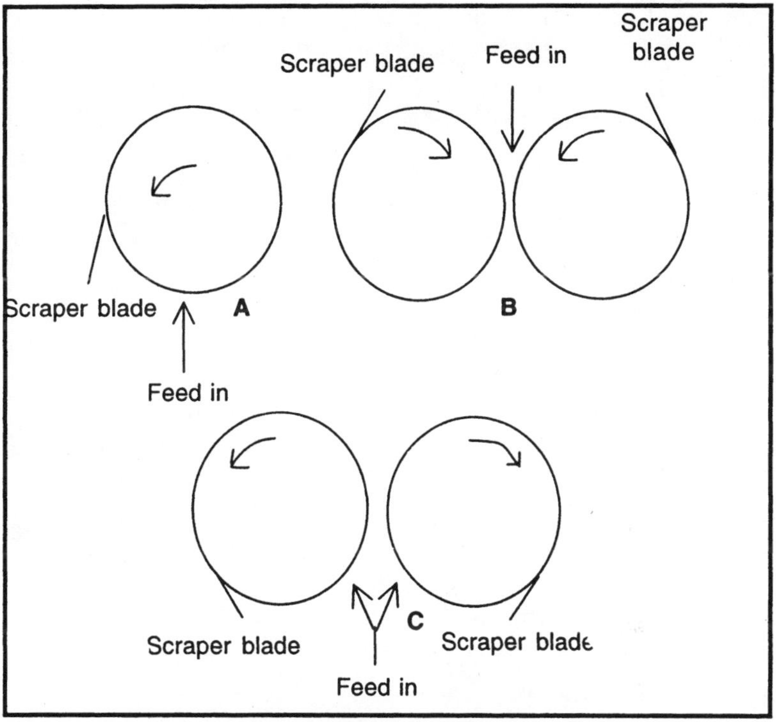

Fig 4.8 : Types of drum driers (A) Single drum, (B) Double drum, (C) Twin drum

4.5.4.2. *Vacuum Shelf Drier*

Vacuum shelf drier consists of a vacuum tight chamber of heavy construction with access door and outlet for gases and vapours. Hollow shelves, through which the heating medium is circulated, are fitted inside the chamber. The material to be dried is spread in fairly thin layers in metal trays which rest on these shelves. Alternately the material can be spread directly on the shelves. Heating is done by circulating hot oil, steam or any other suitable heating medium. Vacuum will be drawn in the chamber through the vapour outlet, and drying will proceed under vacuum. The initial drying rate will be high; however, as the drying proceeds the material will shrink and tend to curve away from the trays. This will reduce the effective area of contact of the material with the heating surface which will cause a decline of heat transfer to the material thus slowing down the drying rate.

Vacuum drying is considered an expensive process. However, it is quite suitable for drying fatty fishes where the probability of fat oxidation and rancidity in the product can be minimised.

4.6 SOLAR DRYING

Harressing the solar radiation for solar powered driers has, of late, attracted considerable interest because of the absence of any energy cost and possibility of producing a dry fish in good hygienic condition even when the RH is high. Energy of the sun is collected and concentrated to produce elevated temperatures suitable for drying several commodities including fish. When the temperature of the air is raised, its RH will be reduced, in other terms its capacity to hold water will increase and hence, can absorb additional quantities of vapour.

4.6.1 Solar Tent Drier

One of the simplest forms of driers to use solar energy for drying fish is the solar tent drier. This is working on the principle that a black surface absorbs sun's energy far more effectively than any light coloured surface. The air thus heated is allowed to pass through the fish and escape out through a vent in the top simultaneously admitting fresh air inside through a vent provided at the bottom of the tent. Schematic drawing of a polythene tent drier is presented in Fig. 4.9.

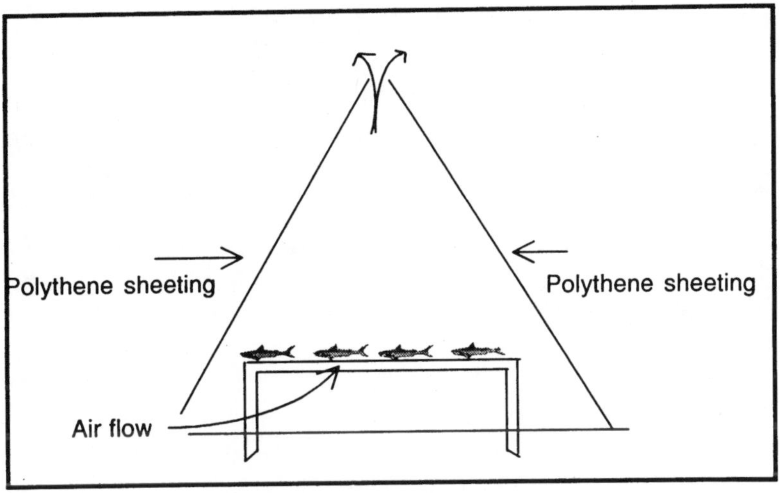

Fig. 4.9 : Polythene tent drier

In a solar tent drier, the air temperature is known to rise to the levels of 60°C or more in tropical climate. This will adversely affect the nutritional as well as physical properties of fish and is considered a disadvantage. However, compared to normal sun drying solar drying has the following advantages :

- no energy cost
- very low equipment cost
- shorter drying periods
- no contamination from dust, insects
- produces hygienic product with low moisture content

4.6.1.1 Other Types of Solar Driers

The tent drier described in section 4.6.1 above is perhaps the simplest in design and operation. Several attempts have been made to improve the efficiency of the drying operation, and accordingly, several designs have come up. Many of these have not become popular due to some reason or other. However, solar drying for some other commodities like food grains has become a commercial success.

4.7 EFFECT OF DRYING ON THE QUALITY OF FISH

4.7.1 Shrinkage

The major apparent physical change which can be observed in dried fish is the shrinkage in its volume. Shrinkage takes place as a result of the structural changes taking place in the fish muscle during drying. Ordinary drying is liquid phase drying and when water leaves its place, proportionate shrinkage in volume of fish also takes place.

4.7.2 Case Hardening

Water in the fish contains dissolved salts, proteins and other organic matters. Water moving to the surface of the fish carries all these, while that leaving the fish surface is only pure water depositing the dissolved substances on the surface. If the temperature of drying air is high and its RH is low this will form a dry impervious layer on the surface preventing further diffusion of water. This condition is referred to as 'case hardening'. When case hardening takes place, the temperature of the fish muscle will increase resulting in its cooking. This will result in the final product becoming brittle. Case hardening greatly affects the rehydration property of dried fish very adversely.

Case hardening can be controlled by maintaining sufficiently high RH in the drying atmosphere as also controlling the temperature of drying.

4.7.3 Denaturation of Protein and Toughening of Texture

As the drying progresses, concentration of the dissolved material in the body water increases. Reduced evaporation due to case hardening will result in increase of temperature of the fish muscle. These bring about denaturation of protein and the texture becomes tough besides loss of juiciness of the meat.

4.7.4 Rehydration

Air-dried fish is hard in texture. The proteins are denatured and the water-holding capacity of the proteins become irreversibly lost. Therefore, penetration of water will be greatly retarded resulting in poor rehydration.

4.7.5 Effect on Colour, Flavour

Pigments, fat etc. in fish are susceptible to oxidation. Prolonged drying of fish exposed to hot air can accelerate this process. Therefore air-dried fish often suffers from discoloration and rancid odours and flavours. Another type of discolouration is the non-enzymatic browning developed through Maillard reaction. Free sugars react with free amino groups and produce brown coloured products in fish.

4.8 SPOILAGE OF FISH DURING DRYING AND STORAGE

Dried and drying fish are susceptible to many types of spoilage which can affect the quality and shelf life. Damages occurring due to flies and insects are of great significance in open drying under the sun. These are, to a great extent, controlled in mechanical drying, but the post-process infection as well as damages taking place during storage are of equal importance in either type of drying. Most of these can be controlled by adopting hygienic practices in the production side as also maintaining appropriate storage atmosphere.

4.8.1 Moulds

Moulds can grow on the salted or unsalted dried fish if the moisture content is high and high RH above 75 per cent prevails in the storage. Moulds are likely to grow if the temperature of storage is maintained in the range 30-35°C. With the onset of moulds the surface moisture may increase and the fish may become susceptible to other types of spoilage.

4.8.2 Insect Infestation

Unsalted dried fish are often infested with blowflies, *Chrysomya* spp., *Lucilia* spp., *Sarcophaga* spp. etc. Adult flies are attracted to the wet fish and their larvae feed on it. Small fish may dry quickly and hence the larvae may perish before highly damaging the fish. However, larger fish may remain wet for longer periods and, therefore, the incidence of damage due to flies will be more in large fish.

Infestation with flies can be largely reduced by maintaining hygienic conditions in fish handling and processing premises, eliminating areas where larvae can pupate and adult flies lay eggs, and using salt in the drying process.

Insect attack takes place in dried fish during storage as well. Beetle infestation is a serious problem in dry fish during storage. *Dermestes* spp. are the predominant blow flies associated with this. They can grow at moisture levels around 15 per cent. Their larvae feed on the fish, often leaving only bones.

Although salting does not result in killing the larvae or adult flies, presence of salt can reduce their activity.

4.8.3 Rancidity

Fatty fish, particularly, are prone to oxidation and development of consequent rancid flavour. Though some degree of rancid flavour may be acceptable in dried fish, excessive rancidity will be objectionable. Rancidity can be controlled to some extent with airtight packaging of dried fish.

4.9 PACKAGING

Dry fish poses some problems with respect to their packaging needs.

- the irregular shape, whether dried whole or as pieces;
- sharp protrusions which may puncture the packaging material and occupy space;

- dried fish often may be brittle and will break up when packed in a regular shaped package;
- individual packaging for fish will shoot up the cost and
- dried fish, if not appropriately protected by package will absorb moisture and spoil quickly, necessitating water-proof/vapour-proof packaging

An ideal package for dry fish should have the following desirable properties :

- be resistant to mechanical abrasion and puncture. To reduce the breakage during handling and distribution, the package should be sufficiently robust to retain its integrity. Containers of corrugated fibreboard will be good as they can provide adequate mechanical protection and
- be impermeable to moisture, oxygen, and insects. Boxes made of corrugated fibreboard or polypropylene can meet these requirements if they are also sealed throughout

However, it remains a fact that practically there is no proven solution to all the technical problems in packaging of dried fish. Any improvement in the existing system should be appropriate to the system, particularly in economic terms.

Suggested Reading

Cutting, C.L., Reay, G.A. and Shewan, J.M. (1956) *Dehydration of Fish*. HMSO, London.

FAO (1983) *The Production and Storage of Dried Fish*. FAO Fisheries Report No.279 (supplement) FAO, Rome.

Fennema, R.O. (1982) *Principles of Food Preservation Part II. Physical Principles of Food Preservation*. Marcell and Dekker Inc., New York.

Heid, J.L. and Joslyn, M.A. (1981) *Fundamentals of Food Processing Operations*. The AVI Publishing Company, Westport, Connecticut, USA.

Jason, A.C. (1965) Drying and Dehydration. In *Fish as Food* Vol. III, Borgstrom, G. (ed.) Academic Press INC, New York.

Seow, C.C.(ed.) (1988) *Food Preservation by Moisture Control.* Elsevier Applied Sciences, London and New York.

Troller, A.J. and Christian, J.H.B. (1978) *Water Activity and Food.* Academic Press Inc., New York.

Van Arsdel, W.B. and Copley, M.J. (1963) *Food Dehydration* (Vol. I), The AVI Publishing Company, Westport, Connecticut, USA.

Van Arsdel, W.B. and Copley, M.J. (1964) *Food Dehydration* (Vol. II), The AVI Publishing Company, Westport, Connecticut, USA.

Waterman, J.J. (1976) *The Production of Dried Fish.* FAO Fisheries Technical Paper No.160. FAO, Rome.

5
SALT CURING

SALTING is a traditional method of preservation of fish, practised as such or in combination with drying or smoking, and is considered a practice as old as drying. Salt curing, though is an important method of preservation by itself, with or without subsequent drying is one of the most widely practised methods of fish preservation throughout the world. When introduced in sufficient quantities in the fish flesh, salt can delay the activity of bacteria or even inactivate them by reducing the water activity. This forms the basis of preservation by salting.

The advantages of salt curing over many other methods of processing are :

- does not require elaborate equipment;
- capital outlay is small;
- methods are simple and the processes are comparatively inexpensive;
- unlike other sophisticated methods of processing, curing can be applied for preservation of any type of fish;
- the finished products do not require any special storage facility;
- the products have reasonably good shelf life and
- nutritionally the products are comparable to fish processed employing most other methods.

5.1 SALTING AND RELATED WATER ACTIVITY (a_w) OF THE FISH FLESH

When fish is mixed with salt or kept in salt solution, some water is removed from the flesh depending on the amount of salt used, dry or in solution. Loss of water from fish flesh reduces its a_w. During salting process salt enters the fish flesh and water moves out due to osmosis. The method of salting is so selected that the salt penetration is rapid enough to similarly lower the water activity in the deepest part of the fish. This process will continue until an equilibrium is attained between the concentration of salt in the fish muscle and the surrounding brine. Water activity of saturated brine is 0.75. Fish flesh in equilibrium with saturated brine also will have a_w similar to this, say 0.75.

5.1.1 Salt Concentration and the Bacterial Growth

Presence of salt at 4-10 per cent level in the fish flesh is known to prevent the action of most spoilage bacteria as well as autolytic decomposition. In general, growth of many putrefactive bacteria is considered controlled with salt concentration over five per cent. However, there are several other bacterial groups which can tolerate higher concentrations of salt. Therefore, for fish curing higher concentrations of salt are always used. When the concentration of salt is 20% or more in the flesh the decomposition process in the fish proceeds only very slowly.

5.1.1.1 Salt Tolerance and Bacterial Groups

With reference to the sensitivity towards salt, bacteria can be divided into three groups

• HALOPHOBIC OR SALT SENSITIVE

These include most of the pathogens and putrefactive types; e.g. *Pseudomonas* and *Achromobacter* spp. These organisms cannot grow in a medium where the salt concentration is higher than six per cent.

• HALOTOLERANT

These consist of spore-formers, micrococci and some anaerobes, particularly *Clostridium botulinum*. They can grow in concentrations higher than six per cent and even up to saturation, although the rate of growth decreases with increasing concentration of salt.

• HALOPHILIC

These are salt loving organisms which grow best in the presence of salt. Practically, the halophiles fail to grow in the absence of salt and the optimum growth occurs in salt concentrations higher than 20 per cent. Bacteria causing 'red' and 'pink' in salted fish are the most important in this group. These are aerobes with optimum temperature of growth in the range 15-50°C. Most of them fail to grow below 10°C. They are not pathogenic.

5.1.2 Action of Salt

During salting, sodium chloride penetrates the fish tissues. While doing so, salt alters the colloidal properties of protein and changes the nature of hydration. Sodium chloride being strong electrolyte releases some of the bound water of the protein and this affects its state of existence. In low concentrations, salting actually results in swelling of the flesh whereas at high concentrations salting out takes place whereby the fish flesh is depleted of considerable amount of water. This will partially account for the hard texture of salt cured fish.

Sodium chloride exerts a high osmotic pressure giving rise to reduction in moisture content of the fish and plasmolysis in the bacterial cells as a consequence. It also blocks the protein nuclei, which are affected by enzymes. It alters the state of proteins and enzymes in such a way that the protein becomes impervious to the action of the enzymes. Enzymes are also destroyed or made inactive by the salt, thereby preventing autolytic degradation.

Sodium chloride forms brine using the water on the surface of the fish and penetrates the fish flesh by a dialysis mechanism. Water diffuses from the fish flesh tissues to outside due to osmotic pressure between brine and the fish muscle solutions. The process will continue until equilibrium attains between the salt concentrations in the fish flesh and the surrounding brine.

5.2 SALTING PROCESS

Salting, though a method of preservation by itself, is also resorted to as a preliminary to other methods of processing like smoking, canning etc. The salted fish may be expected to remain stable for a long time, though often it is intended only as a short-term preservation method. Therefore, the selection of method of salting depends primarily on the end product desired.

5.2.1 Preparation of Fish for Salting

Whatever be the ultimate product intended, for successful salting it is essential to ensure that the whole surface of the fish is in contact with salt. To achieve this the fish is dressed. Usually, the gills and entrails are removed and the fish is split open along the vertebral column. In the case of big fish, deep scores are also made at several places in the split fish, or they are filleted. However, when very small fish are salted, they are salted whole. Principally, it is the size of the fish that decides whether the fish should be salted whole, uneviscerated, eviscerated, split open or made into fillets.

5.2.2 Salting Methods

Dry salting, kench salting, brine salting and, mixed salting are the common methods employed for salting fish.

5.2.2.1 Dry Salting

Dry salting is the simplest method of curing fish. Dressed fish are kept intimately mixed with crystalline salt in containers. Salt may even be rubbed into the gill cavities and the scores made in the fish flesh. A layer of salt is spread on the bottom of the tub and a layer of fish is placed on it skin down. Salt and fish are spread in alternate layers, the proportion of salt increasing upwards. A solution of salt will be formed in the water exuded from the fish and the fish will remain in this self-brine. The fish may float in the brine formed. In order to ensure proper salting of all the fish, weights are often used to keep them immersed in the brine. Fish will be allowed to remain in brine for two to three days after which it can be taken out and dried or otherwise disposed.

5.2.2.2 Kench Salting

Kench salting is essentially a method of dry salting except that the self-brine formed is allowed to drain off. Split fish, after rolling up in crystalline salt or rubbing salt onto the surface, is stacked in layers interspersed with a thin layer of salt between each layer to a height of 1 to 1.3 m. In the lower layers the fish is stacked flesh up and in the top layer it is kept skin up. The self-brine formed is allowed to drain away.

5.2.2.3 Brine Salting

In this method of salting, fish is kept immersed in brine of desired concentration for the required time. This is usually done as a step preliminary to smoking and canning.

A strong cure is not possible in brine salting because the brine will become progressively diluted with the water released from the fish. If a strong cure is needed the brine should be replaced with a strong one after the initial brine becomes diluted.

5.2.2.4 Mixed Salting

Salting process is slow when the fish is large in size or is oily. As a result, it takes long for the self-brine to form and cover the fish. This may cause the fish in top layers to spoil before salt is picked up by them. An advantageous method of salting employed in such cases is mixed salting, also called pickle curing.

In mixed salting, particularly used for medium size fatty fish like herring, the fish is first mixed with dry salt and then packed in watertight containers with salt sprinkled between each layer. It is then topped with saturated brine. A cover is placed on top of the fish to hold them below the surface of the brine.

Mixed salting has the advantage that the fish are completely surrounded by brine right from the beginning, thus quickening the salting process.

5.3 FACTORS AFFECTING SALTING PROCESS AND QUALITY OF THE PRODUCT

5.3.1 Source, Composition and Quality of Salt

5.3.1.1 Source

Pure common salt is sodium chloride. Commercial common salts may contain certain impurities, their levels depending on the source of salt and the method of production. The most common types of salt used are solar salt and rock salt.

• SOLAR SALT

Solar salt is produced by evaporating sea water or salt-lake water under the sun. Solar salt usually contains several impurities including a multiplicity of salts other than sodium chloride. Solar salt is quite often contaminated with sand and also may harbour high numbers of halophilic bacteria.

• ROCK SALT

Rock salt is mined from natural deposits. This generally contains varying percentages of sodium chloride in the range 80-99 per cent. Rock salt is ground to varying degrees of fineness before use.

5.3.1.2 *Chemical Composition*

Suitability of the salt for fish curing will depend on its chemical composition, microbial purity and physical characteristics.

Commercial salts contain sodium chloride ranging from 80-99 per cent. They may contain chlorides and sulphates of calcium and magnesium, sulphates and carbonates of sodium, and traces of copper and iron as chemical impurities. Ca^{++} and Mg^{++} ions from impure salt bind with proteins and form a barrier to the Na^{+} thus delaying the penetration of salt into the fish. An increase in the hydration of proteins and coagulation of gel occurs which are salted out of solutions. Magnesium chloride is hygroscopic and tends to absorb water. This will make it difficult to keep the fish dry.

Even a small quantity of 1% Ca^{++} + Mg^{++} in salt imparts a white colour to the fish. However, even this small quantity imparts a bitter taste as well. Presence of 0.5 per cent of calcium and magnesium as sulphate in the salt can impart a desirable whiteness and rigidity to the fish flesh, but higher concentrations will lead to excessive bitterness and brittleness.

Multivalent ions of copper and iron catalyse the development of brown or yellow discoloration in salted fish. Even 30 ppm iron and 0.2-0.4 ppm copper in salt catalyses browning and rancidity of fish. This will impart an undesirable appearance to the fish even though the eating quality is not affected.

Salt used in fish curing should have at least 95 per cent sodium chloride. It should also be free from halophilic bacteria and as dry as possible. Calcium and magnesium impurities can be washed away using clean water as they dissolve quicker than sodium chloride and can be removed along with wash water. This should be resorted to only if purer common salt is not available.

5.3.1.3 *Microbial Quality*

Most of the solar salts have been shown to contain a sizeable population of halophiles. One of the halophilic groups can cause a red or pink discoloration on wet or partially dried salted fish and hence is of great commercial significance. They do not grow if the fish are fully immersed in brine or fully dried. Halophilic moulds are met with more frequently in rock salt.

5.3.1.4 *Crystal Size*

The size of salt crystals greatly influences the salting time. Coarse crystals dissolve only slowly and take more time to attain the desired concentration of brine. With large and coarse crystals the area of contact with fish will be less, adversely affecting its dissolution and salting. Too fine crystals may cause too much dehydration on the surface of contact with the fish leading to the consequent hardening and thus delaying further penetration of salt. This condition is called 'salt burn'. The size of salt crystals chosen shall be one that will ensure a faster rate of dissolution.

5.3.2 Nature of the Fish

Frozen and thawed fish allow greater permeability of salt into the flesh. The fresher the fish, the slower will be the diffusion of salt to the centre. Fish in rigor takes longer time for salt uptake compared to fish in which the rigor is resolved and is in the early stages of spoilage. Another factor determining salt uptake is the fat content. Higher amount of fat in the flesh acts as an effective barrier to salt uptake and slow down the process significantly.

5.3.2.1 *Fish Size and Dressing*

Salting time can be considered as reciprocal to its specific area. A flat fish with a large specific area will take less time for salting. Salting process can be greatly accelerated by splitting open the fish, thus providing greater specific area per unit weight for the salt to come into contact with. Still greater area of contact can be made available if deep scores are made in the split fish flesh. This will further accelerate the salting process. Thickness of the fish is another factor deciding the salt penetration. More the thickness of the fish, the slower will be the rate of diffusion of salt to the centre.

Skin and scales are considered the greatest barriers to salt penetration to the fish flesh. However, skinning or scaling of individual fishes to increase the salt penetration is not a practical one. Very small fishes are salted as such; still bigger ones may be eviscerated and their gills removed before salting.

5.3.3 Salt Concentration in Brine

The rate of diffusion of salt into the fish flesh and that of water from the flesh tissues to the surroundings will depend upon the concentration gradient of salt in the two media. Greater the concentration of the salt in the surrounding brine, faster will be its uptake by the fish flesh. However, it need not always be strictly proportional to the concentration gradient, as different salt concentrations have been found to give rise to different changes in the protein ranging from hydration to salting out.

5.3.4 Temperature

Salt penetration into the fish flesh increases with increase in temperature. Increase in temperature intensifies the thermal motion of particles and also causes a reduction in the viscosity of water. But increase in temperature also may increase the rate of spoilage. If, however, the fish is salted at lower temperatures, the rate of penetration will be reduced; but the rate of spoilage will be reduced even more. Uptake of salt by fish fillets at two different temperatures is given in Table 5.1.

Table 5.1 : Percentage uptake of Salt by Fish Fillets at Two Different Temperature

Temperature °C	Percentage uptake of salt		
	1 day	2 days	3 days
5	1.1	4	14.1
27	4.8	8.6	15.2

5.4 TYPES OF SALTED FISH

Based on the quantity of salt used, salted fish can be divided into three categories - light salted, medium salted and, heavily salted.

Salting aims at saturating the water in the fish flesh wholly or partly with common salt. Salted fish can be saturated or unsaturated with salt depending on the quantity of salt used. The maximum concentration of salt attainable in the fish flesh is equivalent to that of saturated brine, i.e., around 2[illegible] per cent. In practice, it will be lower because of the presence of other dissolved substances in the fish body water. Assuming that fish muscle contains approximately 80 per cent moisture, about 20-22 kg salt will be required to saturate 100 kg of fish with salt. In practice 30-34 kg of salt is used for 100 kg of fish to get a heavily salted fish. The resultant product will contain 40-45 per cent salt on dry weight basis. If a smaller quantity of salt is used, the cure will be unsaturated.

For light salted fish 16-20 parts of salt per 100 parts of fish is used. The resultant product will contain around 20-30 per cent salt on dry weight basis. For medium salted fish 20-28 parts by weight of salt is used for every 100 parts fish and the resultant product will contain 30-40 per cent salt on dry weight basis.

In processing light salted fish, the duration of salting is of great significance. Too long a period of salting will lead to onset of spoilage of fish during the salting time itself. Drying or other types of further processing will have to be carried out immediately. Brown discoloration and rancidity due to fat oxidation and other putrefactive changes are quicker and more pronounced in light salted fish. Salting time generally does not exceed 24 hours for light salted fish.

5.5 MATURATION IN SALTED FISH

Maturation is a phenomenon observed in brine pickled fatty fish during storage. During maturation the fish lose their raw flavour. The weight lost through loss of water by osmosis is regained through salt uptake within a few days and the fish flesh becomes tender and acquires a special pleasant taste. Maturing is considered to take place by the action of enzymes, proteolytic and lipolytic in that order. The enzymes responsible for maturation are derived from the digestive system of the fish, primarily from the pyloric caeca.

Maturation is also dependent on the bacteria present in the fish as well as in the brine. Carbohydrates are fermented and provide the matured fish a pleasant flavour and slight sour taste.

The process of maturation is a highly complex one. The products of proteolysis and lipolysis are seen predominant in the product although several others like fermentation products of carbohydrates, the products of Maillard reaction etc. also contribute to the odour and flavour of matured fish.

5.6 SPOILAGE OF SALTED FISH

Bacteria exhibit different degrees of tolerance towards salt. Though salt is known to have bacteriostatic/bactericidal action, depending on its actual content many bacteria may become active. Most of the bacteria usually associated with fish spoilage are halophobic in nature and will not grow when the salt concentration is in excess of 5 per cent. However, there are other organisms which are halophilic and can grow in environments of high salt concentrations. In addition to the action of halophiles changes occurring in the protein, fat etc. also will contribute to the spoilage of salted fish.

5.6.1 Microbial Spoilage

5.6.1.1 'Pink' or 'Red'

This is a common type of spoilage associated with salted fish manifesting mostly during storage in warm weather conditions. The surface of the fish becomes covered with a red slime that gives off an unpleasant odour. This is brought about by halophiles which need salt concentration above 10 per cent for their growth. The spoilage is called 'pink' or 'red' because of the colour of the colonies of the bacteria appearing on the fish surface.

The microorganisms responsible for this phenomenon are halophilic rods and cocci originating from the salt, mainly solar salt used in the process. They include *Halobacterium salinaria, H. Cutirubum, Sarcina morrhuae, S. Littoralis* and *Micrococcus rosens.* These are all aerobic organisms active only while in contact with air. They are thermophiles with an optimum temperature of growth of about 42°C. They will not grow at temperatures below 10°C.

The spoilage manifests first as a delicate pink sheen on the surface of the fish. At this stage it can be rubbed or washed off easily without damage to the fish. Fish will remain edible as the bacteria responsible do not produce toxin. However, if severely infected by the bacteria the red will reappear and the surface of the fish will become softened due to protein decomposition giving off ammoniacal odour. The flesh turns alkaline and becomes inedible.

The bacteria responsible being aerobic and thermophilic in nature, an ideal preventive measure will be to keep the fish out of contact with air and store at low temperatures, preferably below 10°C. After initial occurrence and washing off the slime, the recurrence of 'pink' can be prevented by treatment with formaldehyde or sulphur dioxide vapours or by dipping in a solution of sodium metabisulphite. Treatment with sodium/calcium salts of propionic acid also effectively controls the development of 'pink'.

5.6.1.2 *Moulds*

Moulds are often seen to grow on salted and unsalted dried fish. Moulds will grow profusely if the RH is 75 per cent or more. They are also temperature sensitive having an optimum temperature for growth of 30-35°C. 'Dun' is a type of mould development observed even in heavily salted fish. This is characterised by the appearance of coloured spots, black, grey or brown, visible particularly on the fleshy side of the fish. This imparts an appearance as if sprinkled with ground black pepper. The small spots develop a root-like network into the interior of the fish flesh. This is caused by a group of moulds, of which *Sporendonema epizoum* is mostly identified with. Some strains of *S. epizoum* are mildly proteolytic, but does not cause softening of tissues. However, the discoloration is unsightly and reduces the marketability of the product. *S. epizoum* is a true halophile and will flourish if the storage atmosphere is damp. It has optimum growth at 10-15 per cent salt concentration, 75 per cent RH and 25°C. The mould activity itself may make the surface moist and pave the way for the fish becoming more susceptible to other types of spoilage.

The moulds can be relatively easily removed in the early stages. They can be easily scraped or brushed away; however, they will reappear rapidly. In cases of severe infection dipping in 0.1 per cent sorbic acid will give some protection.

Solar salt is known as one of the sources of contamination; effective control may be achieved by use of good quality salt, maintenance of low temperature and humidity and, well ventilated and dry storage conditions.

5.6.1.3 *Saponification*

The spoilage referred to as saponification is a damage caused by aerobic microorganisms active even at very low temperatures and is identified by a malodorous slime on the surface of the fish. This phenomenon is associated with light salted fish of the type, boxed herring, when in contact with air. In the initial stages of its development, the fish remains edible if washed with water and treated with a solution containing vinegar and water.

To prevent saponification the fish can be kept in brine containing vinegar for some time before boxing.

5.6.1.4 Putrefactive Spoilage

If the salting process is very slow, it will take unduly long time for the salt to reach the centre of the fish and to saturate it. This can set in motion a putrefactive process in such regions. The flesh near the backbone become 'tanned' or reddened accompanied by the development of an offensive putrid smell. Tanning usually appears near the tail and kidney. Any presalting operation which can accelerate the penetration of salt to the interior of the flesh like gutting, splitting etc. can prevent the development of tanning.

5.6.2 Non-microbial Spoilage

5.6.2.1 Infestation by Maggots

Infestation by flies is a very common and serious problem faced by the salt fish trade. Infestation actually takes place during the early stages of drying. Cheese flies (*Drosophila casei*) are attracted to the drying fish by its smell and the unhygienic atmosphere in the drying premises due to the presence of discarded and spoiled fish and other materials. Adult female flies lay their eggs on the fish and on the sides of the containers. The eggs hatch in a day or two and the larvae (maggots) jumps about and spread rapidly and soon infest the whole fish causing considerable damage to the fish. The larvae metamorphose into red pupae from which fully-grown flies emerge. The emerging adults can then reinfest more fish and the cycle will continue.

Fish infested with maggots can be salvaged to a great extent if immediately washed thoroughly in brine and placed in clean uncontaminated containers. It is advisable to plunge the whole lot of fish into the brine when the maggots will float at the top. They can then be collected and destroyed. If fresh water is used for washing they will sink to the bottom and can be drawn away.

Infestation by cheese flies can be controlled in several ways. The best choice is minimising the chances of their coming into contact with the fish and fish handling premises. This include maintaining the handling and processing premises clean and fly-proof, ensuring that the fish is well covered during salting with no access for the flies to get in and lay the eggs etc. As the cheese flies thrive in warm weather and multiplies freely at temperatures above 20°C, it is desirable to maintain the handling and processing premises cool as a control measure.

5.6.2.2 *Rust*

Appearance of a colour similar to that of rusted iron on its surface is a phenomenon most commonly associated with salted fish. Oily fishes like sardine, mackerel etc. are particularly prone to rusting. Rusted fish is characterised by, besides its colour, an unpleasant taste and rancid odour. These are brought about by oxidation of oil in the fish by the atmospheric oxygen. The natural antioxidants in the tissues of live fish disintegrate soon after its death and hence the body fat becomes devoid of any protection. Further, the salt present in the salted fish accelerates the oxidation process. Fat is first oxidised to the hydroperoxides at which stage this cannot be detected organoleptically. However, the chain of reaction continues and the hydroperoxides break down to aldehydes, ketones, hydroxylic acids etc. These products are responsible for the rancid odours and flavours as well as the discoloration. Traces of iron and copper in the curing salt can catalyse the oxidation and consequent discoloration considerably.

Rust first appears on the surface and then penetrates the skin and spread throughout the flesh making it inedible. When it first appears on the surface, the impairment of the flavour is not significant. It can be washed away using dilute solution of sodium bicarbonate.

The best method to prevent the occurrence of rust in salted fish is to prevent it from having any contact with air. Fish should be kept covered with brine during salting and storage. Dry fish should be kept properly packed so as to avoid contact with air to the extent possible.

5.6.2.3 *White spots*

It has been mentioned (Section 5.3.1.2) that presence of calcium and magnesium in the salt can produce remarkable whitening in the flesh of salted fish. Different from this, some white spots occur occasionally on the surface of salted fish which consist of crystals of disodium hydrogen phosphate derived probably from the enzymatic breakdown of nucloetides. The causes for the occurrence of white spots have been described as :

- partial drying of salt bulk prior to storage;
- low temperature of storage and
- initial spoilage of fish prior to salting.

Exposure to dry air will partially dehydrate the crystals leaving them as a scarcely noticeable white powder.

5.6.2.4 *Fragmentation*

Cured and dried fish often become brittle and break during storage and transportation. This is referred to as fragmentation. Denaturation of proteins, hollowing of the fish by insect attacks, use of spoiled fish for processing etc. are the reasons ascribed to this. Fragmentation can be reduced by using fresh fish for processing and providing adequate protection against physical damage by using appropriate packaging.

Suggested Reading

Burgess, G.H., Cutting, C.L., Lovern, J.J. and Watermann, J.J. (eds) (1967). *Fish Handling and Processing*. Chemical Publishing Co. Inc., New York.

Cutting, C.L. (1996) *Fish Processing and Preservation.* Agro Botanical Publishers, India.

FAO (1982) *Prevention of Losses in Cured Fish.* FAO Fisheries Technical Paper No.219. FAO, Rome.

Seow, C.C.(ed.). (1988) *Food Preservation by Moisture Control.* Elsevier Applied Science, London and New York.

Watermann, J.J. (1976) *The Production of Dried Fish.* FAO Fisheries Technical Paper No. 160 FAO, Rome.

6
SMOKING

SMOKING or smoke curing, like drying and salt curing, is an ancient method of preservation of fish. In the early methods of smoking, heavily salted fish used to be smoked for long durations, few weeks even, and the resultant products were called 'hard cures'. These products had long shelf life at ambient temperatures owing to the high salt concentration and long smoking and drying periods which lower the water activity considerably. In course of time, the 'hard cures' gave way to milder products with less salt and lower durations of smoking. Such products, however, have only short shelf life. These are more relished because of their specific mild flavour and are considered delicacies. Smoking is also employed as an intermediate step in processing canned smoked fish. Lightly smoked fish is considered as an alternate to fresh fish having slight pleasant smoke flavour.

6.1 FUEL AND DECOMPOSITION PRODUCTS

6.1.1 Fuel

Fuel is the source of smoke as well as of heat. Smoking should impart, in addition to the specific aroma and taste, an attractive colour to the product. Fuel selected for generating smoke should be such that it meets the above requirements. Wood is the most commonly used fuel and therefore selection of the type of wood for fuel is a very important consideration in the smoking process. Wood is used in the form of sawdust, shavings or logs. It is also important that the wood does not impart any unpleasant flavour or colour to the product. Hard wood species are preferred as fuel for smoking. Soft wood species like conifers are not suitable because they contain volatile resins, which will impart undesirable flavour to the product.

6.1.2 Composition of Wood

Wood contains both combustible and non-combustible substances, the latter being constituted by water and ash. The combustible components are complex organic substances consisting mainly of polyoses (including pentosans) and lignin. Soft wood species will contain resins also in addition to the above. Combination of combustion and destructive distillation occurring while smoke is produced from smouldering wood results in a complex mixture of aliphatic and aromatic compounds in addition to water, carbondioxide and traces of hydrogen and carbonmonoxide.

6.1.3 Physical and Chemical Characteristics of Smoke

6.1.3.1 Physical Properties

On heating, wood undergoes destructive distillation and releases gases and vapours in the temperature range 100-150°C. The vapour consists mainly of water. The volatile matter released in this temperature range constitutes only about 2 per cent of the total matter released. On raising the temperature to 200°C wood begins to char with an increased output of volatiles upto 25 per cent. Between 150 and 200°C the volatiles generated are mostly constituted by carbon monoxide and carbondioxide. Above 250°C the output of volatiles increases sharply and the quantum of hydrogen and hydrocarbons decreases. At about 300°C wood catches fire.

Smoke is a typical aerosol with the solid and liquid particles dispersed in a disperse gaseous medium. The gases produced during the process of destructive distillation condense in the cool zone above the fire to form a stable aerosol. The disperse medium is a mixture of gases like oxygen, hydrogen, nitrogen, carbon dioxide, carbon monoxide and various hydrocarbons. Many organic substances in the form of tarry droplets are present as vapour or as liquid depending on the source of smoke and conditions of smoke generation. The particle size, which also depends on the conditions of smoke generation and cooling, varies from 0.1 to 0.14 μ in radius.

6.1.3.2 Chemical Composition of Smoke

• PRODUCTS FROM POLYOSES

Cellulose, the cellular fibre constituting the bulk of wood, is a polysaccharide $(C_6H_{10}O_5)_n$ having molecular weight approximately 150,000. At high temperatures it is hydrolysed to simple saccharoses, mainly glucose.

$$(C_6H_{10}O_5)_n + n\ H_2O \rightarrow n\ C_6H_{12}O_6$$

Glucose undergoes further decomposition yielding oxymethyl furfural

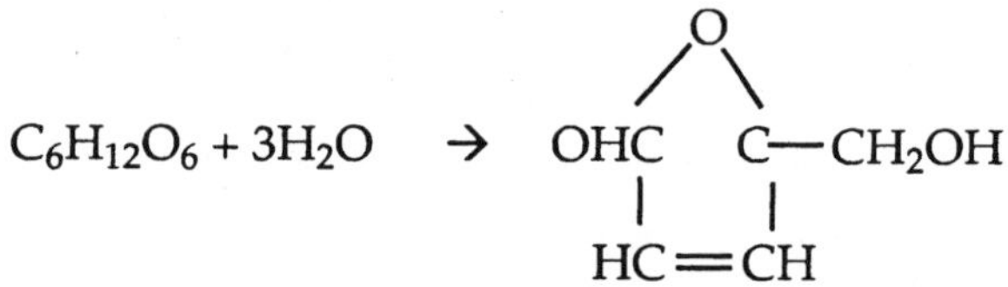

Oxymethyl furfural is unstable and further breaks up yielding formic acid (HCOOH), levulinic acid (CH_3-CO-CH_2-CH_2-COOH) and humic substances which give the smoked fish its characteristic colour. Products of pyrolysis of polyoses can be summarised as

Alcohols	-	methanol, ethanol, propanol
Aldehydes	-	formaldehyde, acetaldehyde, propionaldehyde, arylaldehyde, furfuraldehyde, 5-methyl furfuraldehyde
Ketones	-	acetone, diacetone
Acids	-	formic acid, acetic acid, propionic acid

• PRODUCTS FROM LIGNIN

Lignin is a constituent of the cell walls in the wood. Lignin is more resistant to heat and begins to break down at 350°C. Lignin contains methoxyl group CH_3O- and hence one of the products obtained on pyrolysis is methanol. Other products of pyrolysis of lignin are a complex mixture of phenols and tar. The phenolic constituents consist mainly of guaiacol and its 4-methyl and propyl homologues, catechol and its 4- methyl derivative, pyrogallol and its methyl, dimethyl ethers and homologues and, hydroquinone, phenol and p. cresol.

6.1.3.3 Tar Distillate

Tar distillate contains both low and high-density tar oils. Low density tar oils having boiling points below 140°C contain valeric aldehyde and furanes - furane, methyl furane, dimethyl furane and trimethyl furane. Tar contains guaiacol, vinyl guaiacol, cresol, catechol, phenol, hydroquinones, eugenol and similar compounds.

High density tar oils have boiling points above 200°C and contain phenols and their derivatives like 0-, m- and p-cresols, xylenols, ethyl phenol, cetechol, guaiacol, esters of pyrogallol and lignoceric acid. The aromatic substances in the tar are products of lignin decomposition.

6.1.4 Products of Complete Combustion

When combustion is incomplete, smoke produced contains organic substances as above which can impart a distinct smoky flavour to the fish. But when the combustion is complete organic products break down yielding carbon dioxide and water with no formation of smoke. Therefore when the fuel, which is often saw dust, is burned, the temperature and supply of air are to be so controlled that the fuel burns incompletely and adequate smoke is produced.

6.1.5 Benzopyrene in Smoke

Wood smoke may contain traces of a polynuclear

aromatic hydrocarbon 3, 4 benzopyrene ($C_{20}H_{10}$) which is a carcinogen that can cause tumours in humans and animals. Its quantum may vary depending upon the method of smoke generation. Stronger draught and high combustion temperature are two factors contributing to its increase. However, no harmful effect has been experienced due to this, presumably due to its extremely low concentration. It is also probable that benzopyrene, if present, would tend to concentrate in the particle phase and would not be absorbed by the fish.

6.2 DEPOSITION OF SMOKE

As the disperse medium of smoke is not very viscous, intense Brownian movement of the dispersed solid and liquid particles occurs. The colloidal particles collide, coagulate and form flakes. Smoke is stabilised by polar compounds like phenols, amyl alcohol etc. in the surface vapours. With maximum stabilisation the adsorption layer is completely saturated. Further concentration of stabiliser results in partial coagulation of aerosol. Water is a polar substance and with high relative humidity the smoke will coagulate with tarry substances settling on the surface of the fish.

During smoking largest concentration of smoke is seen at the top of the kiln. Accordingly, deposition of tarry substances will be more on fish at the top than on those in the lower layers.

6.3 PRESERVATIVE ACTION OF SMOKE

Preservation by smoking is due to the combined action of salting, drying and deposition of natural chemicals produced during the thermal breakdown of wood. The roles played by individual constituents of smoke are not fully understood, still the importance of several components can be explained with reasonable accuracy.

- salting which is an essential step prior to smoking proper, reduces the a_w and inhibits the growth of many spoilage microorganisms as well as pathogens

- smoking causes partial surface drying of the fish. This will present a physical barrier to the passage of the microorganisms and also retard the proliferation of aerobic microflora. Smoking at higher temperatures, as in hot smoking, will destroy the enzymes as well as bacteria.
- phenolic constituents like guaiacol and its homologues are effective antioxidants providing protection against autoxidation of lipids
- antimicrobial agents like phenols, formaldehyde etc. deposited on the fish control the microbial spoilage to a great extent and are considered responsible for the biological stability of smoked fish.

6.3.1 Bactericidal Property of Smoke

The bactericidal properties of smoke is due to the presence of chemicals such as formaldehyde, phenols, organic acids and other organic compounds that make up the tarry constituents of the smoke. Smoke behaves differently towards pure cultures of microorganisms and microflora of fish. Pure cultures of non-sporers like streptococcus are killed by smoke in three hours, whereas those of spore-formers like *Bacillus subtilis* and *B. mesentricus* survive even beyond seven hours exposure to smoke. Wood tar phenols are most active inhibitors of *Staphylococcus aureus.* As regards the effect of smoke on the microflora of smoked fish it is known that smoke below 40°C has only a weak antibacterial action. It is only during hot smoking that the number of microorganisms is considerably reduced, which is mainly due to the high temperature of smoke.

6.3.2 Significance of Phenols

6.3.2.1 *Colour Formation*

One of the indices for deciding the adequacy of smoking fish is its phenol content. The formation of colour in smoked fish is considered related to the deposition and subsequent oxidation of phenols. Oxidation of phenols is accelerated under alkaline conditions.

6.3.2.2 *Antioxidant Effect of Phenols*

Wood smoke is also known to have a pronounced antioxidant effect providing effective protection against autoxidation of fat in fish. This is attributed to guaiacol and its homologues. 1, 3 dimethyl pyrogallcol also contributes to the antioxidant property to a lesser extent. Highly active antioxidants catechol, 4 methyl catechol and pryogallol 1-0-methyl ether from the particle phase of electrostatic smoking provide greater protection against autoxidation.

6.4 GENERATION OF SMOKE

6.4.1 Smoke from Burning Wood

Wood is the source of smoke used the world over. Wood in the form of chips or coarse shavings or as saw dust is most commonly used. Wood chips and sawdust is an ideal combination. A layer of wood chips or coarse wood shavings with a layer of sawdust atop is used for generation of smoke. As air cannot get easy access to fire, the sawdust smoulders rather than burns. Lower temperature and limited supply of oxygen causes production of smoke with high content of flavouring and preserving principles.

6.4.2 Frictional Smoke Generator

Another method of production of smoke is using frictional smoke generators. In this process the end of a piece of wood is pressed against a rapidly revolving drum or disc. Due to friction the temperature of wood rises up to 350°C and smoke is generated. Frictional smoke is generated at a temperature lower than that of charring smoke. Frictional smoke contains approximately four times the quantity of volatile acids, eight times the quantity of volatile aldehydes and ketones and three times the quantity of phenols compared to smoke from charring wood.

6.5 PREPARATION OF FISH FOR SMOKING

6.5.1 Splitting, Gutting etc.

Quality of the raw fish has significant influence on the quality of the smoked product. Pelagic fishes are considered best for smoking when their fat content is the highest; however, depending on the end product desired, fish of different fat contents are preferred. Generally fish for smoking should be kept chilled. In certain cases of non-fatty fishes it is considered desirable to keep the fish iced for two days before smoking.

Fish may be smoked whole or after splitting, gutting etc., the method employed being dependent on the end product desired, e.g., smoked herring is processed in split and ungutted form. In some products the fish is split with the backbone retained on one fillet. Large fish are gutted, headed and split or made into fillets. It is very important that the guts, gills, kidney etc. are completely removed. These are the areas where spoilage can take place quickly and can affect the overall quality of the end product, particularly the light smoked fish which are more perishable because of the low salt content and light smoking at low temperature.

6.5.2 Salting

Fish is salted either by mixing with solid salt or by immersing in strong brine. Size of the fish, fat content, whether the fish is whole, gutted, split or cut, presence or absence of skin and other requirements in the end product etc. are the factors considered in determining the duration and type of salting. Depending upon the concentration of brine there may be an uptake or loss of water in the fish. However, there will be no appreciable difference if the saturation of the brine used is 70-80 per cent and for this reason, this is the most common concentration of brine employed. Fish salted in 100 per cent saturated brine is not preferred as it is likely to make the product unattractive because of the powdery, unattractive, crystals of salt that may be left on the surface. In weaker brines longer periods of immersion are needed and there will be an uptake of water which will have to be evaporated during drying. Salting is so adjusted that the fish takes up 2-3 per cent salt which is the optimum requirement in many smoked fish. However, when the fish is dry salted the uptake of salt will be very high. Such fish are soaked for several hours in water before smoking to bring down the salt content to the desired level.

6.5.3 Hanging

After salting the flesh of the fish will be moist. Surface proteins will dissolve in the brine yielding a sticky solution. For efficient smoking fish must be dry enough with no free water on the surface. Free water can cause condensation of the distillation products, particularly tar, which leave dark brown colour in the product. During hanging the sticky solution dries on the surface and makes the skin glossy. However, excessive drying will prevent the fish from acquiring a proper smoky flavour. Drying can be done in open air or a hot air drier or even directly in the smoke chamber by burning wood without producing smoke.

Split and salted fish should remain in a stretched out position for uniform exposure of all parts to smoke. Hanging is done in several styles depending on the fish handled and the end product desired. Fish like sardine meant for canning after smoking are threaded through the eyes on thin speats. Large split whole fish like salmon are suspended by the tail and kept flat open by means of sticks or skewers threaded through the flesh. Fish can be hung or even spread on wire mesh trays, however, care has to be taken to ensure that the mesh does not leave excessive mark on the skin or flesh.

6.6 SMOKING PROCESS

Two types of smoking are popular, cold smoking and hot smoking. The essential difference between the two processes is the amount of heat the fish receives during the process. Cold smoking is a mild process where temperature is not raised enough for the fish flesh to undergo even a partial cooking. In hot smoking the temperature of smoke is high and the fish flesh is cooked even to the extent of partial sterilisation. However, this partial sterilisation does not result in any increased shelf life as the fish may get reinfected with microorganisms during subsequent handling and packaging.

6.6.1 Cold Smoking

Cold smoking is still carried out in more or less the conventional style, mostly using traditional chimney kilns. Fish is kept hung or spread in mesh trays in an upward draft of smoke produced in the floor by the burning fire. In the initial stages the fish is moist and the smoke is highly humid. Under these conditions use of high temperature will invariably result in cooking of the fish flesh. To avoid this, the temperature inside the chamber should not be raised to the maximum employed in the process. In the second stage, when the surface of the fish is considerably dry, the temperature inside the chamber could be raised to a level that the fish species concerned will tolerate. However, it is desirable to complete the process at the same temperature throughout. The highest output of good quality products is achieved when the temperature, relative humidity and the quantity of ventilated smoke are maintained in the correct proportion. Fig. 6.1 shows the diagram of a traditional chimney kiln.

During smoking the fish dries as also absorbs the aromatic substances from the smoke. The relative humidity inside the chamber also is a decisive factor in determining the quality and output of the products. RH above 70-80 per cent will considerably slow down the drying and smoking process. To control the RH of the smoke, the flame dampers may be kept open thus providing a good draught whenever the per cent RH goes beyond the required level. Cold

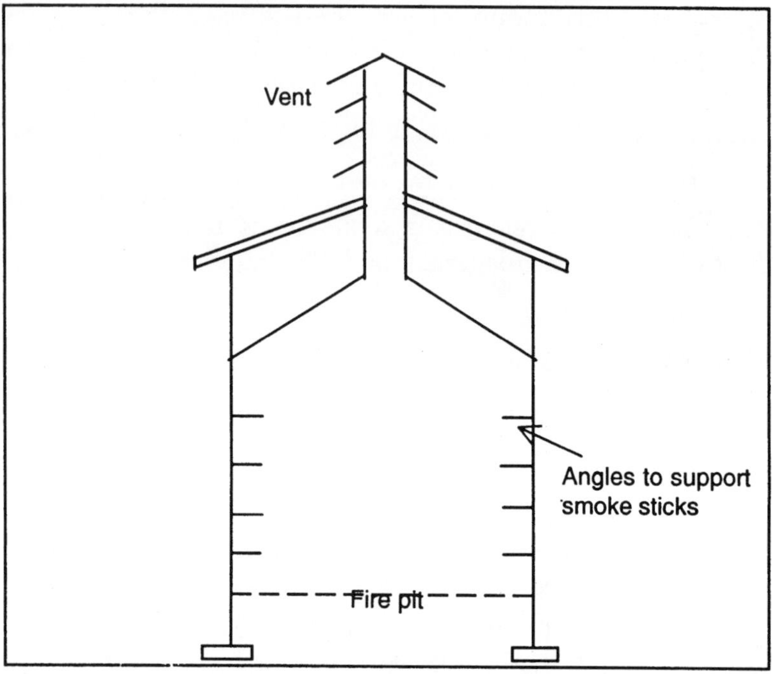

Fig. 6.1 : Conventional smoke kiln

smoking is generally carried out at temperature not exceeding 40°C. Duration of smoking extends to several hours, say 36-72.

6.6.2 Hot Smoking

Several designs of kilns or smoking tunnels are available for hot smoking fish. The fuel is burnt either directly inside the kiln on movable trolleys or in external hearths located near the tunnel. The fish is charged into the tunnel in cages. The chamber will be having a metal frame structure and brick walls and can hold a number of fish cages.

The 'Torry kiln' designed and developed by the Torry Research Station, Aberdeen, is a very popular one and is widely used in the UK for fish smoking. This is a horizontal flow kiln and such kilns are often called mechanical kilns because an electric fan draws the smoke over the fish in racks. Electric heaters are placed at different points in the smoke path to maintain even temperature throughout the smoke chamber. Outline of a mechanical smoke kiln is presented in Fig. 6.2.

In contrast to cold smoking where cooking of the flesh is neither the aim nor is attained in any case, in hot smoking fish is dried and cooked in the kiln before it is smoked. Drying is done in an intense draught of hot air at 75 to 80°C produced by burning fire. The skin of the fish becomes dry while the flesh becomes cooked as well.

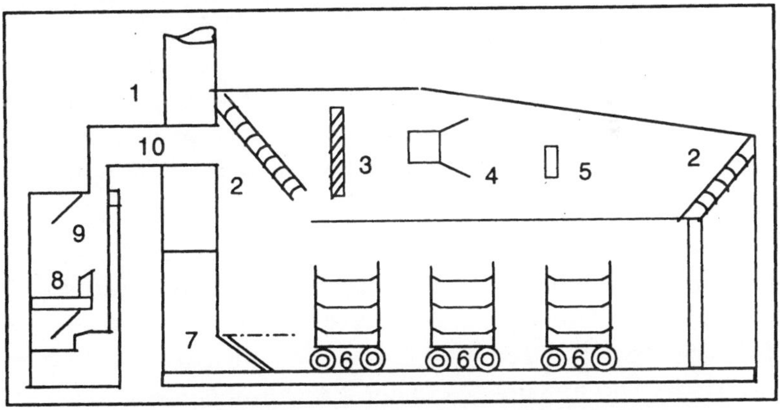

(1) smoke discharge pipe; (2) aerofoil; (3) electric heater; (4) fan; (5) thermostat; (6) fish trolleys; (7) recirculation damper; (8) smoke generator; (9) adjustable flap; (10) smoke duct

Fig. 6.2 : Mechanical smoke kiln

At this stage the fish is considered ready for smoking. Smoke is produced by covering the burning logs with sawdust and the temperature in the chamber is maintained at or above 100°C. A schedule of operation in hot smoking can be considered as drying at 75 to 80°C for an hour followed by smoking at 95 to 100°C for another one hour. The requirements will vary with the size and species of fish.

6.6.3 Electrostatic Smoking

A process of smoking developed and popular in the Soviet Union is electrostatic smoking. Smoking takes place as a result of the electrokinetic properties of smoke in a high voltage field of the order of 40 kW or more. Salted and rinsed fish is passed through a drying chamber heated by infrared lamps positioned on either side of the belt carrying fish. Fish is heated for 3 to 4 minutes at 40 to 50°C in the chamber when they lose around five per cent of its weight. They are then taken to the electrostatic smoking oven over a conveyor. On either side of the conveyor are Nichrome electrodes suitably spaced. The bases of the electrodes are fitted with high capacity heating elements to prevent condensation of moisture. The electrodes are supplied with high voltage current at 30 to 70 kW while smoke is admitted to the bottom of the oven. For an appealing smoked appearance of the fish at least one mg of smoke substances must be precipitated per square centimetre of fish surface. This requirement is attained in less than 5 minutes in the oven.

From the smoking chamber the fish are carried to the baking oven fitted with sufficient number of emitters of required capacity. In 5 to 6 minutes in the baking oven the fish are heated to about 30°C and lose 10 to 12 per cent of moisture. The advantages of the process include

- considerable saving in time as the whole process takes only around 20 minutes
- consequent increased output
- reduction in the losses because of the short processing time
- the process is continuous and carried out in a mechanical equipment

6.6.4 Use of Smoke Concentrates

Smoking fluids or 'smoke concentrates' derived from dry distillation of wood are sometimes used to impart smoked colour and flavour to fish. The salted fish after washing and passing under an air blast is given a short dip of few seconds in the smoking fluid. The fish is then dried in a mixture of smoke and air. During drying the fluid thickens and polymerises coating the fish surface with a thin film with typical golden yellow colour. The process cut shorts the smoking time, increases the productivity and results in considerable saving in materials and outlay. Improved antioxidant effect is considered another advantage. However, the product is rated organoleptically inferior to fish smoked in the normal way.

6.7 USE OF ARTIFICAL COLOURING

It is often difficult to achieve uniformity in the product, particularly colour, in smoked fish. Even if near uniformity in colour can be achieved, batch to batch variation occurs. One way of overcoming this inherent deficiency of smoking process is to artificially colour the fish with permitted dyestuffs. Artificial colouring is done by incorporating suitable dyestuffs in the brine used for salting. Vegetable dyes, coal tar dyes etc are so used, the selection being dependent on the type of the product.

6.8 POST-PROCESS HANDLING

On removing from the kiln, the smoked fish is warm and is generally allowed to cool before grading and packing. Cold smoked fish may lose upto 5 per cent of its weight by further evaporation. If the smoked fish is packed while warm, moisture will condense and deposit inside the containers. This will create a situation conducive to premature mould growth. Smoked fish can also be frozen and cold stored which will minimise the chances of spoilage before consumption.

Suggested Reading

Burgess, G.H.O. and Bannerman, A.Mc.K. (1985) *Fish Smoking*. HMSO, Edinburgh.

Cutting, C.L. (1953) *Fish Saving*. Leonard Hill Books Ltd., UK

Cutting, C.L. (1996) *Fish Processing and Preservation*. Agro Botanical Publishers, India.

FAO (1970) *Smoke Curing of Fish*. FAO Fisheries Report No.88, FAO, Rome.

FAO (1971) *Equipment and methods for improved smoke drying of fish in the Tropics*. FAO Technical Paper No.104, FAO, Rome.

Gopakumar, K. (1997) *Tropical Fisheries Products*. Oxford & IBH Publishing Company, New Delhi.

7
MARINADES

MARINADES are, by general definition, fish or fish portions processed by treatment with edible acids and salt, and put up in brine, sauce or oil. Adjuvants for seasoning and bleaching are also often used. Fish marinades are characterised by the typical marinated odour and flavour. Treatment with acetic acid and salt brings about a sort of refinement in the fish simultaneously developing the typical marinated odour and flavour. Further improvement in odour and flavour is brought about by addition of various spices and covering liquids. Pelagic fatty fish such as sardine, herring etc. are generally used to process marinades, though marinades are also processed out of several other species of fish and shellfish. In order that the characteristic odour, flavour and taste of marinades are maintained, use of additives which can suppress spoilage and prolong shelf life is restricted. Therefore, marinades have only a limited shelf life and are classified among the semi-conserves.

7.1 PRESERVATIVE ACTION OF ACETIC ACID AND SALT

7.1.1 Action of Acetic Acid

Development of putrefactive microorganisms is greatly delayed in an acetic acid medium having acid concentration of 1-2 per cent. If the concentration of acid is higher, many putrefactive bacteria will perish. It is only some microorganisms like moulds and yeast that will thrive in an acid medium. Moulds can gradually decompose acetic acid thereby creating a favourable atmosphere for the development of putrefactive bacteria. On the positive side, the characteristic succulence and tenderness of the marinated fish flesh is produced by the action of acetic acid which favours the action of some of the proteolytic enzymes. The action of these enzymes is responsible for the partial breakdown of proteins and release of some free amino acids which impart the characteristic flavour and taste to the marinated products. Acetic acid in the concentrations employed lower the pH of the meat to the required level creating an atmosphere for the autolytic reaction to take place. At higher concentrations, these reactions will be arrested and hence only lower concentration of 1-2 per cent acid is used.

7.1.2 Action of Salt

Salt also has different functions to perform in a marinade. Salt extracts water out of the fish tissues and causes the proteins to coagulate allowing the proteolysis to proceed at the desired level. Further, salt and acetic acid act in a mutually complementary manner. For example, yeast develops very rapidly in an acetic acid medium; however, stops growing at a 14 per cent salt concentration when the pH is maintained at 2.5. An equivalent effect is obtainable in a natural medium only when the salt concentration is 20 per cent or above. Thus the combined action of acetic acid and salt can favour employment of lower concentrations of both.

7.2 TYPES OF MARINADES

On the basis of the process employed the marinades are classified into three groups

- cold marinades
- cooked marinades
- fried marinades

7.2.1 Cold Marinades

As the name indicates, cold marinades are a class of products processed in the cold. The fish and other ingredients are not subjected to any heat treatment at any stage in the processing schedule. Both processing and storage are carried out at temperatures less than the ambient, say 10-12°C.

Cold marinades are the marinades proper and are processed mostly out of fish like herring, sprat and sardine.

7.2.1.1 Preparation of Cold Marinades

The raw fish is first washed in a solution containing 8-10 per cent salt for about an hour to remove the slime and to harden the meat. It is then headed, gutted, filleted or deboned as required. The fish or fish portions are again placed in a bath containing 3 -5 per cent salt for about 30 minutes. This will facilitate removal of traces of blood from the muscle. A small degree of bleaching desirable in the product can be achieved by adding 0.5 per cent acetic acid to the salt bath. Following this pre-treatment the product is introduced into the finishing bath which is also called the softening bath.

The finishing bath consists of a solution of salt and acetic acid, their strength being determined by the ratio of the fish to solution and the type of the end product desired. Two important chemical processes take place in the fish in the finishing process. One is the reaction of acetic acid producing the characteristic tenderness in the marinade which is largely due to the proteolysis induced by autolytic tissue enzymes, and the butter-like taste by the liberation of amino acids. The other is the action of salt causing removal of water and coagulation of proteins. The salt also inhibits hydrolysis and causes it to proceed within desired limits. The development of aroma is ascribed to microbial action. Therefore the ratio of fish to pickle is so adjusted that at the end of the process the finishing bath has an acetic acid concentration of at least 2.5 per cent. The general practice is to use salt 1-7 per cent higher than that of acetic acid depending on the strength of the final bath. A usual strength of the solution used is 7 per cent acetic acid and 14 per cent salt. If the fish is held at chill temperatures, the strength can be less, however, the concentration must be increased during warmer seasons of the year to prevent spoilage. If the acid and sal levels are too high, the microorganisms cannot survive and the characteristic flavours may not develop.

Pickling is done at an optimum temperature of 10-12°C. Higher temperature will favour the process proceed more rapidly, but the flesh will become too soft and its succulence will be lost. At lower temperatures development of the characteristic flavours is retarded. Pickling is generally concluded in 3-7 days. Pickling is carried out in wooden vats or barrels, open or closed. In any case the container should be full and have a tight fitting lid to prevent contact with air and the consequent development of rancid flavours.

The fish at the end of the process should be firm, white, opaque and tender. It should not be glassy or shining in the innermost portion of the flesh and should be free from any red spots. The marinated fish are packed often in glass jars, and covered with a solution containing 1-2 per cent acetic acid and 2-4 per cent salt.

The acid taste of the cold marinades may be reduced by using citric or tartaric acid replacing some or all of the acetic acid. However, the pH of the final solution should be maintained at 4.5. To improve the flavour and soften the acetic acid taste, the acetic acid-salt solution is often flavoured with spices like pepper, clove, coriander, bay leaves etc. Use of spice extracts is considered better as they, in general, will be free of bacteria. Slices of onion, carrot, celery etc. may serve as garnish.

Shelf life of cold marinades may be several months at chilled temperatures, but in tropical conditions it may be only a few weeks.

7.2.2 Cooked Marinades

Cooked marinades are generally packed in a jelly and are also called jellied products. The fish or fish pieces are treated with edible acids and salt and are heat treated. The combined effect of heat and acid in the bleaching bath is considered the preservative principle in the cooked marinades. Low pH is often maintained in cooked marinades to prevent the growth of harmful bacteria like *Clostridium botulinum*. The shelf life of cooked marinades can be upto six months.

7.2.2.1 Preparation of Cooked Marinades

The fish is dressed and salted as for cold marinades. After draining off the brine the fish is spread on perforated trays and immersed in a bleaching bath containing 1-2 per cent acetic acid and 6-8 per cent salt maintained at 85°C for 10-15 minutes. During this process fish suffers a weight loss of about 15-20 per cent in the form of water removed from it. This will dilute the bleaching bath and hence it must be renewed from time to time. At the end of immersion the trays are taken out of the acid bath and rinsed free of adhering fat and protein foam. Product is packed in appropriate containers made of glass, porcelain or lacquered cans. The concentration of acetic acid in the jellied brine should be about two percent, and the salt content about three percent. Generally a 4-5 per cent gelatin is considered appropriate in the jellied brine. In preparing jellied brine the temperature should not exceed 70°C as gelatin is very heat sensitive. Spices and saccharin can be added to the jellied brine. The brine is cooled to 4°C before filling.

7.2.3 Fried Marinades

In the fried marinades the fish after pre-treatment is baked or broiled in oil with or without breading and is covered with acetic acid or sauces. The higher temperature of frying kills most of the bacteria in the fish flesh. If stored at 0-8°C fried marinades can have a storage life of upto one year. Fried marinades are made out of different fish.

7.2.3.1 Preparation of Fried Marinades

The fish is dressed and salted as for cold marinade; but usually with the backbone retained. After the fish or fish pieces are drained well, they are treated twice in a breading mixture of equal parts of wheat and rye flour. Leaving for an hour after first breading the fish is again given a second coat of breading. Any excess flour is gently tapped off.

The breaded fish is fried for 5-12 minutes in oil maintained at 160-180°C. The frying oil consists of a mixture of oil and solid fat to obtain a pliable film of fat all around the fish. The duration of frying depends on the temperature of frying oil, thickness of fish and its water content. In a deep fat frying system the termination of the frying process is indicated by the rise of the fish to the surface, because the specific gravity of the fish is altered by the absorption of fat and loss of water. During frying the interior of the fish flesh attain temperature below the boiling point of water, though higher temperature persists in the oil medium. The breaded outer layer attains high temperature, and the flesh takes on a characteristic dark colour and special aromatic components of fried products are formed.

The fried fish is cooled well and packed in cans. It is then covered with brine containing 3-5 per cent salt and 2-3.5 per cent acetic acid. Use of malt vinegar or vinegar containing spices and herbs is recommended for fried marinades.

The ratio of fish to covering liquid is about 2 : 1. Much of the water lost by the fish during frying will be taken up from the covering liquid which will take place gradually. Therefore, the filled cans are allowed to stand open for several hours and are then again filled to the brim and then sealed.

7.3 SHELF LIFE OF MARINATED FISH

Marinades in general are stable only for a limited period. The only preservatives in cold marinades are acetic acid and salt. In cooked and fried marinades, even though there is a heat treatment involved, the temperature attained by the flesh is below 100°C which is not sufficient to kill all microorganisms. There is also the possibility of post-process contamination up to the time the cans are sealed. The shelf stability depends mainly on the storage temperature and, to some extent, on the type of bacteria present. Spoilage in cold marinades is indicated by gas formation; gas formation and liquefaction in jelled products, and gas formation with appearance of slime in fried marinades.

Spoilage occurring in fish marinades is classified into physical, chemical and biological types.

7.3.1 Physical Spoilage

If the pack of marinades is stored below 0°C the contents will freeze. Water expands on freezing and may damage the glass jar or tin can. In cooked marinades the structure of jelly will be damaged by the formation of ice crystals. Heat damage will occur if the cans are filled when the product is too cold, and stored in warm environment. Cans may swell if cooked marinades are sealed in cans before swelling of the flesh is completed. However, the contents of the can are very rarely altered by physical spoilage.

7.3.2 Chemical Spoilage

Chemical spoilage is caused by the action of acetic acid on the metal of the badly lacquered and/or tinned cans. Because of the corrosive action of the acid on the metal, hydrogen gas will be liberated which will cause the cans to swell. Metal dissolved in the product will also impart a metallic taste to the product and alter its flavour making the product unacceptable.

7.3.3 Biological Spoilage

Biological spoilage may result in simple flipping to hard swells and even in bursting of the cans. Biological spoilage in fish marinades is of two categories, one due to the microbial degradation of proteins in the fish and the other due to the fermentation of the added sugar or other sugar containing additives. Biological spoilage also results in production of gas, mainly carbondioxide and also in the degradation of the quality of the product.

Distinction between chemical and biological spoilage can be made by analysis of gas, bacteriological and chemical investigations and also by organoleptic evaluation.

The principal causes of spoilage in marinated fish are heterofermentative lactic acid bacteria in cold marinades, proteolytic bacteria in cooked marinades and slime formers in fried marinades. High content of acetic acid and salt can control the action of putrefactive bacteria and slime formers. Suppression of lactic acid bacteria is very difficult, as it will require very high concentrations of acetic acid. Fish marinades are distinguished for their taste and aroma and conditions required to arrest the bacterial action, that is use of high concentrations of acid and salt, will impair both taste and aroma. Therefore the most important requisite in processing marinades is high grade raw materials supplemented with maintenance of strict hygiene and sanitation in the premises and processing schedules including work tables, utensils etc. throughout the process.

7.4 OTHER TYPES OF MARINATED FISH

There are certain other products, which are considered as marinades though they do not strictly come under this group of products. The South American specialties called *Ceviche* and *Escabeche,* and *Paksiw* of the Philippines, are prominent among them.

Ceviche is made by marinating fish or shellfish with sour orange juice, lime juice or tartaric acid dissolved in water. To process *ceviche* skinless fillets of fish is diced into 10-15 cm cubes, combined with a mixture of onion and garlic, and seasoned with salt. The marinating solution, viz. the orange/

lime juice or tartaric acid solution, is poured over this and let to marinade over night. The product has a shelf life of less than five days in tropical temperatures.

For making *escabeche,* fish pieces are fried and marinated in vinegar. Vinegar may, often, be flavoured with garlic, ginger and pepper. Sugar also may be added.

For *paksiw,* raw fish together with coconut, vinegar, salt and sometimes sugar are boiled together in a pot. Water is added and the mixture is re-boiled. The product has a shelf life of several days.

Suggested Reading

Clucas, I.J. and Ward, A.R. (1996). *Post-harvest Fisheries Development: A Guide to Handling, Preservation, Processing and Quality.* Chatham Maritime, Kent, ME4 4TB, United Kingdom.

Davidson, A. (1976) *Seafood of South East Asia.* Macmillan, London.

Gopakumar, K. (1997) *Tropical Fisheries Products.* Oxford & IBH Publishing Co. New Delhi.

Mayer, V. (1965) *Marinades.* In *Fish as Food,* Borgstrom, G. (ed.) Vol. III Academic Press Inc. New York.

8
FERMENTED PRODUCTS

FERMENTATION is an age-old technology of preservation of highly perishable freshwater and marine animals. However, processing of fermented fish and fishery products is almost exclusively confined to the Asian region. Only very few of these products are known to be processed outside Asia.

Most, if not all, processes employed in fish preservation are so designed that the identity of the fish is unaltered, and aim at keeping the fish meat as near as possible to its original state. If the fish is adequately protected against contamination by microorganisms and endogenic enzymes, the flesh will not break down and remain remarkably stable for a considerable period. However, fermentation involves breakdown of proteins in the raw fish to simpler substances which are themselves stable at normal temperatures of storage. Fermentation is generally described as a process in which the complex protein molecules in the fish are broken down by the action of organic catalysts, enzymes or ferments, into simpler molecules. In some processes where the process is controlled by adding salt, only a partial breakdown of the proteins takes place. Such processes are so controlled that a product with a desired type of flavour is produced simultaneously ensuring preservation of the product. Breakdown of protein can be brought about by the action of exogenic as well as endogenic enzymes, the latter being present in the guts and intestines of fish. Sometimes microorganisms are also involved. Fermented fish products have a meaty flavour and provide dietary needs of amino acids, some vitamins and other nutrients.

8.1 FERMENTED FISH PRODUCTS

The original fermented fish product was perhaps, the liquid exuded from fish during salting. Fermentation of fish without addition of salt, but employing other agents, is also popular. Different processes employed in the fermentation of fish yield three distinct types of products as follows :

- products in which the fish substantially retains its original form
- products in which the original fish is reduced to the form of a paste and
- fish sauces in which the fish meat is reduced to a liquid.

Fermented fishery products are generally classified into three categories, high-salt (20 per cent or more salt), low-slat (6-18 per cent salt) and no-salt products.

High-salt (20 per cent or more)	- Liquid separated → fish sauce
	- Residue → cured fish
	- Mincing and partial drying → fish paste
Low-salt (6-18 per cent)	- Lactic acid fermentation
	- Acid pickling at low temperature
No-salt	- Dried bonito fermentation
	- Alkaline fermentation

Though most mammalian enzymes are highly proteolytic and active at low pH, the visceral and digestive tract enzymes in the fish are active at or near neutral pH. Various cereals and plants are added in many cases so that the digestive enzymes from these sources also aid in the process of fermentation.

Fish subjected to salting alone is likely to ferment to some degree. The degree of fermentation will depend on the proportion of salt used, fat content of the fish, dressing of the fish like complete or partial removal of the gut, the nature of the additives, if any, added and the temperature at which the salted fish is maintained. The temperature is of particular importance. This has been evidenced by the observation that using precisely the same process, but by maintaining at different temperatures, products quite different in characteristics can be obtained.

With salt concentration higher than 20 per cent three different types of fermented products can be obtained. If fermentation is allowed to complete, i.e., when all the protein matter has been broken down to simpler water soluble compounds, the resultant liquid is called fish sauce. When it is partial, the fish is likely to substantially retain the original form in which it was treated with salt. The resultant product is cured fish. This product is minced and partially dried or otherwise processed to yield fish paste.

When the concentration of salt is less than 20 per cent as in the case of low-salt process, the fish may undergo spoilage due to the action of microorganisms. Therefore, some means of preservation other than the use of salt is also essential. In a low-salt process, lactic acid fermentation along with added sources of carbohydrates is a common practice. Rice, milk, flour and syrup or sugar is used as the source of carbohydrates. Carbohydrate and salt control the extent of fermentation as well as the keeping quality.

An alternative method is the acid pickling and storage at low temperature. This involves keeping the low-salt fermented fish with added vinegar at low temperatures, practised in the Scandinavian countries.

Fermentation without salt is not a common practice. However, alkali fermentation and fermentation by propagation of mould are practiced as some local specialties. In the alkali fermentation, often half spoiled fish is fermented using leafy plant ash as the source of alkali. Processing of bonito by propagation of mould on the dried fish is another example of no-salt fermentation.

Another classification of fermented products is based on the technique employed in the process. Two major groups are (i) products primarily involving hydrolysis by enzymes and (ii) the product preserved by microbial fermentation. It is true that the bacteria may rapidly die off in the fermentation media containing salt; however, those capable of surviving in such media are also capable of supplying both proteolytic and lipolytic enzymes which can aid in the enzymatic fermentation process. It is also known that the microorganisms capable of growing in the medium contribute to the development of the characteristic aroma and flavour.

8.2 LIQUID FERMENTED PRODUCTS (SAUCES)

Unlike many other techniques employed in fish preservation, no standard procedures have been laid down for fermented fish products. The type of raw materials, processing schedule, additives employed etc. will vary depending on the local tastes and preferences. The major concern in the marketability of fish sauce is the branded aroma developed by blending sauces made out of different fish.

8.2.1 Processing

In the production of fish sauce, fish and salt mixture is allowed to stand for extended periods until the whole flesh is converted into a liquid containing amino acids and other protein breakdown products that are water-soluble. Fish of all types and even marginally spoiled fish can be used in this process. Methods of processing some of the traditional products are described below.

8.2.1.1 Nuoc-mam

The method of processing fish sauce varies with different countries. *Nuoc-mam* is a traditional product of Vietnam and is the favourite in Southeast Asia. *Nuoc-mam* is predominantly produced out of anchovies (*Stolephorus* spp.) as it is considered to yield one of the best sauces made from fish. An outline of the important steps involved is as follows :

(i) The whole fish is washed and mixed with salt in the ratio varying from 1 : 5 to 1 : 1 arranged in layers in tubs or vats made of wood or cement. Weight is placed on the top to keep the fish below the brine formed in the container. After 3-4 days a liquid containing blood and salt called 'blood pickle' which oozes out is removed and kept aside. The fish is then mixed well and the blood pickle is poured over the fish to form a 10 cm layer over it and the container is kept covered.

(ii) The fish is allowed to remain in this condition for five to 18 months. The clear liquid floating on the surface is the *nuoc-mam* which is siphoned out. The liquid sauce is filtered and packed in bottles or earthen containers and kept exposed to sunlight until it is ripe.

A second harvest is usually possible by adding salt and water freshly to the system. However, the overall quality and the protein content in this product will be poor.

A first grade *nuoc-mam* should contain 20-23 g nitrogen per 1000 ml and 50 per cent of this should be formol titratable and not more than 15-20 per cent titratable as ammonia. The salt content should be 20-25 g/100 g and the pH below 6.0

8.2.1.2 Patis

Patis is a fish sauce produced in the Philippines. Its production process is almost the same as that of *nuoc-mam.*

Small fish are salted whole; however the large ones are gutted and cleaned. They are then chopped to pieces of convenient size. Varying ratios of salt to fish are employed depending on fish quality. Fermentation is carried out in drums or cement tanks.

Budu of Malaysia, *nam-pla* of Thailand etc. are some other important traditional fish sauces.

8.3 FACTORS AFFECTING PROCESSING AND QUALITY OF FISH SAUCE

8.3.1 Salt

Though the growth of putrefying bacteria is prevented by the generally high salt content and the low pH of the medium, halophilic and halo-tolerant bacteria can grow well in a fermenting medium.

Salt content above 20 per cent is required to prevent the excess conversion of organic nitrogenous compounds to ammonia. Such sauces will contain only 10 per cent of the total nitrogen as ammoniacal nitrogen. When the salt content is less, this may go to very high levels. In the high-salt fermentation process, proteolysis is mainly carried out by the intestinal enzymes of fish and involvement of microorganisms or enzymes of microbial origin is minimal.

8.3.2 Lipid Content

Fermentation is an anaerobic process. Therefore, oxidation of lipids and the consequent development of oxidised flavours can be considered not very significant. However, marginally spoiled fish which may contain some oxidised fat are often used to process fish sauce. Pelagic fish of high fat content are also very popular raw material for fish sauce. Consequently, presence of some oxidised fat in the sauce is quite natural. The oxidised fat may affect the colour as well as aroma of the sauce.

8.3.3 Colour

Colour of the sauce varies with the fish used in its production. Most of the well-known sauces like *nuoc-mam, nam-pla* and *budu* are principally processed out of anchoviella. The colour of sauce produced from fresh anchoviella varies from lemon yellow to light brown. Those produced from oil sardine (*Sardinella longiceps*) or mackerel (*Rastrelliger kanagurta*) are deep brown to dark red in colour.

8.3.4 Flavour

Perhaps the most important quality criterion of a fish sauce is its typical aroma. Several volatile and non-volatile compounds contribute to the flavour of sauce. Volatile compounds include ammonia, tri- and di-methyl amines, low molecular weight aliphatic carboxylic acids and their esters. Non-volatile compounds are the protein breakdown products like peptides, peptones and amino acids, higher fatty acids and their esters, sugars and their derivatives, products of ATP breakdown etc.

Sauces from different fishes processed by different methods show variations in their aroma. The characteristic aroma of a branded product is often developed by blending sauces from different sources.

8.3.5 Accelerating the Rate of Fermentation

Fish sauce production is a time consuming process and often takes over a year for the process to complete. Many attempts have been made to speed up the process. A method employed is to carry out the process at accelerated temperatures upto 45°C. The fermentation process gets accelerated, however, the resultant sauce becomes unacceptable on account of the absence of the characteristic flavour of the classic sauces.

Another method of acceleration of the fermentation process tried and found partially successful is to carry out the fermentation in the traditional style for some period, say 50 days, and then to continue it under accelerated conditions by artificially heating and aerating the fish/salt mixture in a concrete tank. This process will be complete in two months in place of 12 months for the traditional process.

8.3.6 Nutritional Quality

Though it contains the protein breakdown products like amino acids, peptones etc., fish sauce is not considered a nutritionally important product because its use is limited due to the high salt content and the hazards associated with the intake of high amounts of salt. Fish sauce is principally used as a condiment to flavour and salt many vegetables and animal product dishes.

8.4 FERMENTED FISH

Different from fish sauce, fermented fish is a product in which the identity of the fish is maintained to a large extent. Fermented herring and anchovy are very popular in the west; products locally called *makassar* (Indonesia), *buro* (Philippines) or *pekasam* (Malaysia) are the important products of the East. Colombo-cured fish is a well-known product of India coming under this group.

8.4.1 Methods of Processing

8.4.1.1 Pickled Herring

Pickled herring (*Clupea harengus*) is dry salted with 15-36 per cent salt in barrels. The self-brine formed will cover the fish. After a while, if necessary, the barrels will be topped up with fish and brine of the same day's curing. Some sugar is also added to the salt as an aid in fermentation. Spices like pepper, mace, coriander, hops, cinnamon, ginger and cardamom are often added to improve the flavour. Benzoic acid also is occasionally added as a preservative. Fish pickled in this way can be kept for more than a year at the European ambient temperature. The fish flesh becomes only moderately softened, the ripening process taking several months.

8.4.1.2 Cured Anchovies

This is another delicacy popular in the Mediterranean region and is made out of *Engraulis encrasicolus*. Anchovies with very high fat content and weighing 35-40 g each are considered the best for curing. The fish is nobbed, the guts being pulled out along with the head, and is salted in layers using 5-6 kg salt per each 10 kg fish. The fish is topped with a layer of salt and the fish is maintained under pressure with a weighted wooden disc kept on the top. The fish will sink in the self-brine formed and additional fish and salt will be added a few days later. The fish will be kept in brine under pressure for at least six or seven months. During this period, water and fat pressed out of the fish form a layer of brine covered with fat. The liquid, which overflows, is collected and later used to spray the fish during cure.

Curing is carried out in sterile containers using sterile salt and, therefore, no microorganisms are involved in the process.

8.4.1.3 Colombo Curing

Colombo curing is a pickling process practised in the South Canara and Malabar regions along the West Coast of India. Colombo cured products used to be processed mostly out of mackerel (*Rastrelliger kanagurta*). The fish is first gutted and washed free of dirt and slime followed by rubbing with salt (fish to salt 3 : 1) and kept in cement tanks. Pieces of *Garcinia cambogea* fruit are mixed with fish/salt mixture at about 8 kg of fruit per tonne of fish. *Garcinia cambogea* fruit is very acidic and hence it imparts acidity to the pickle and makes the product sour. It is the smoked-dried pulp of the fruit that is used for pickling. The fish remains in the tank for 3-4 months.

In another method, a smoked-dried piece of the *Garcinia cambogea* fruit is placed inside the belly flap of each fish. The fish with the fruit and salt are arranged in layers in large wooden vats. The vats are filled upto the top and kept tightly closed upto three months after which the product is marketed.

8.5 PASTE FISHERY PRODUCTS

Fish paste is another product involving fermentation and processed in many Asian countries. The processes employed for making all fermented fish pastes are essentially similar. Fish is mixed with salt and pounded to form a paste. The paste is sun dried for varying periods before it is sealed in containers from which air is excluded for maturing. In some cases, comminution of the fish and salting precedes a period of sun drying. Moisture content of the final product is in the range 30-40 per cent.

Typical among the paste fishery products are *bagoong* of the Philippines, *ngapi* of Myanmar, various *mams* of Vietnam, *belacan* of Malaysia and *trassi* of Indonesia. A typical process for making a fermented paste product from small shrimp practiced in Malaysia is as follows.

The raw material used is usually *Acetes* shrimp together with small proportion of mysid shrimp. The shrimp is mixed well with salt in bamboo baskets or wooden tubs. The proportion of salt to fish is generally 4-5 kg salt to 100 kg wet shrimp. The mixture is then spread out in thin layers on mats and allowed to dry in the sun. Drying is continued until around 50 per cent of the water is evaporated. During drying occasionally it is turned with a shovel to quicken the drying process. Drying may extend upto 4-8 hours. The semi-dried shrimp is minced in a mechanical mincer. The resultant paste is filled in wooden tubs/boxes under pressure to exclude all air and prevent oxidative rancidity. The paste is maintained upto seven days for fermentation under this anaerobic condition. It is then taken out of the box and spread on a mat and dried in the sun for 4-5 hours. The paste is minced again and filled in the wooden tubs and allowed to ferment for about a month after which it is again minced and packed in blocks wrapped in cellophane or laminated paper. If kept under refrigeration *belacan* has a shelf life of several months.

The yield of the paste is 40-50 per cent of the raw shrimp weight. A good quantity *belacan* will have pH 7.6-7.8, moisture 20-40 per cent, ash (including salt) 20-24 per cent, salt 13-18 per cent and protein 30-40 per cent.

8.6 LACTIC ACID FERMENTED PRODUCTS

Fish products fermented with lactic acid are generally processed out of freshwater species of fish. Fermentation with lactic acid is quicker than that with salt alone. Such products will contain lower levels of salt in the range 6-18 per cent compared to over 20 per cent in sauces. Their processing involves mixing the fish with salt and a carbohydrate such as cooked rice.

The keeping quality as well as the extent of acid fermentation will depend on the amount of salt and carbohydrate used. Carbohydrate is used as a food source for the lactic acid producing bacteria.

Pla-ra is a prominent lactic acid fermented product of Thailand. The process in its production involves the following stages.

The fish is scaled, eviscerated and mixed with salt in the ratio 3 : 1 and is then packed in jars and kept for periods extending from 15 to 90 days. After the stipulated storage period the fish is taken out of the jars, washed well and allowed to drain. It is subsequently mixed with ground roasted rice in the ratio 1 : 10 rice to fish and again placed in the jars for further storage for one to six months.

Consistency of *pla-ra* varies from a dark brown liquid to partly dried fish pieces. It may even be eaten as the main course.

Suggested Reading

Adams, M.R., Cooke, R.D. and Pongpen Rattagool (1985) Fermented Fish Products of South East Asia. *Tropical Science,* 25 : 61-73.

Erichsen, I. (1983) *Fermented Fish and Meat Products: The Present Position and Future Prospects.* Academic Press, New York.

FAO (1971) *Fermented Fish Products.* FAO Fisheries Report. No.100, FAO, Rome.

Gopakumar, K. (1997) *Tropical Fishery Products.* Oxford & IBH Publishing Co., New Delhi.

Reilly, P.J.A, Parry, R.W.H. and Barile, L.E. (eds) (1990) *Post-Harvest Technology. Preservation and Quality of Fish in Southeast Asia.* International Foundation for Science, Stockholm.

Rose, A.H. (ed.) (1982) *Fermented Foods.* Academic Press, London and New York.

Steinkraus, K.H. (ed.) (1983) *Handbook of Indigenous Fermented Foods.* Marcel Dekker Inc., New York and Basel.

Van Ven, A.G. (1965) *Fermented and Dried Seafood Products in Southeast Asia* In : *Fish as Food.* Vol.III. Borgstrom, G. (ed.) Academic Press Inc., New York.

9
CANNING

CANNING is a method of preservation of foods in which spoilage is averted by killing the microorganisms present by application of heat and prevention of subsequent contamination, the product being enclosed in a hermetically sealed container. Hermetic sealing literally means air-tight closing by fusion. By definition, canning is the method of food preservation in which selected food materials are prepared for the table, packed in tin or glass containers capable of being sealed airtight, heated sufficiently to destroy the spoilage organisms within the container and cooled rapidly. Unlike other common methods of preservation canning alters the nature of the material significantly forming almost new products because of the various treatments the raw materials are subjected to, and the various additives used in processing. Containers for canned foods are normally made of tinplate, but aluminium and other modifications are now popular.

9.1 THE BEGINNING

All methods of preservation of foods, perhaps with the exception of irradiation, can be considered adaptation of natural processes or their modifications. Canning has the unique distinction of being an invention. The history of development of canning as an important method of preservation of foods dates back to the mid-1790s. In 1795 the French Government, faced with the problem of feeding the fighting forces, announced a prize of '12,000 Francs and Fame' to anyone inventing a useful method of food preservation. In 1809 Nicholas Appert, a French confectioner, won this prize and fame 'benefactor of humanity' for his invention when he found that foods remained safe for long periods if heated in a sealed container. Appert called his method of food preservation 'Appertisation'. Neither he nor the scientists of this period could give a convincing explanation for the apparent success of the process. It was believed that in some "magical and mysterious" way air combined with foodstuff in a sealed container preventing their spoilage. It was in the eighties that a logical explanation for this could be provided when Louis Pasteur reported that it is a microscopic vegetation that grew and spoiled foods under unfavourable conditions.

Appert used glass containers for his "Appertised" products. Since then the canning processes witnessed tremendous development like introduction of newer types of containers and their manufacturing technology, developments in processing equipment and machinery, exploratory research leading to the proper understanding of the role of microorganisms in food spoilage and methods of their control etc., all of which could place the canning industry on a firm scientific footing. Some of the major landmarks in the history of development of canning as a successful commercial process are the following :

- announcement of a 'prize and fame' by the French Government in 1795 for inventing a useful method of food preservation
- Nicholas Appert wins the prize and fame in 1809 for his invention of 'Appertisation'
- beginning of canning of lobster in England in 1817 and its spread to America with the beginning of lobster and salmon canning there
- invention of "tin canister" for processed foods by Peter Durand in 1810 gave great impetus to the growth of commercial canning The present term 'can'

is the abbreviated form of 'canister'. Metal cans were first introduced in United Sates during the 1820s

- until 1840, heat processing of canned foods was done in boiling water. To increase the temperature of boiling water chemicals such as sodium chloride, calcium chloride etc. were added to the water. Use of steam for heat processing became prevalent by the year 1840.
- though canning as a method of food preservation continued to gain popularity, the underlying principles of preservation was not known to anybody. The scientific basis of the preservation of heat processed foods first became available when Louis Pasteur demonstrated in 1860 that the spoilage in foods is caused by a microscopic vegetative organism and that they can be made inactive by heating
- the canning industry had all the while been facing the problem of condemning several lots of products merely because of insufficient heating in the final heat processing stage. An end to this problem came in sight when Andrew Schriver patented a pressure retort for heat processing canned foods in 1874
- introduction of inside coated cans in 1890 by Max Ames was another achievement in canning
- development of "sanitary cans" around 1900 paved the way for mechanisation of the canning process and increasing the production rapidly
- the scientific explanation given by Andrew Russel in 1895 and the first scientific paper by Prescott and Underwood (1897) threw much light into the scientific basis of the canning process and placed it on a firm scientific footing
- the classical work of Esty and Meyer (1920) on heat resistance of micro-organisms and patenting of Aseptic canning by Ball (1936) are very important landmarks in the development of canning in this period

9.1.1 Advantages of Canning Over Other Methods of Preservation

Canning has several, advantages over other methods of preservation some of which are :

- canned foods offer consumer safety;
- canned foods can be stored at room temperature for long periods;
- canned foods are cooked foods and hence instantly available for consumption with little or no further preparation for the table;
- canned foods are concentrated foods with no waste;
- they are protected against re-infection by microorganisms and insects and
- the process is applicable to a wide range of products.

9.2 OUTLINE OF CANNING OPERATIONS

The different steps involved in the conventional canning process are the following :

9.2.1 Selection and Preparation of Fish

Thermal destruction of bacteria being the principle involved in preservation of fish by canning it is very important that the fish used should have only a very low bacterial load. Thermal destruction of bacteria follows a logarithmic pattern. Therefore, higher the initial load of bacteria, longer will be the time required for their elimination. Longer exposure to high temperature will impair the quality of the product. Therefore, fresh uncontaminated fish should be used as the raw material for canning.

Dressing needs will depend on the type of fish and the type of the end product desired. For small fish like sardine this involves heading, gutting, scaling and removal of fins and tail. Bleeding is necessary for fish like tuna which is done immediately after catch. Shrimp is peeled and deveined for canning. Bivalves such as mussel, clam, oyster etc. require a purification process called 'depuration' to reduce the bacterial population as well as the grittiness.

Dressing reduces the bulk of the raw materials making its further handling convenient. The raw materials other than fish used in the processing also should be of prime quality.

9.2.2 Salting/Blanching/Pre-cooking

Dressed fish is generally blanched in cold or hot brine or pre-cooked in steam, the choice being dependent on the fish concerned. During blanching in brine, hot or cold, the fish flesh takes up sufficient salt and its texture gets improved. During heating, as in hot blanching or cooking in steam, the fish flesh releases around 15-30 per cent of the body water. If the fish is packed without removing this much water beforehand by some process it will be released in the can during heat processing. This will render the product unattractive, form water-oil emulsion or dilute the sauce depending on the canning medium used. Internal corrosion of the can material also may take place. Therefore blanching/pre-cooking is carried out to an extent such that no further water will be released from the fish flesh during heat processing. The main functions of this process can be summarised as :

- causes sufficient shrinkage of the fish to enable adequate filling in the cans;
- imparts firm and proper texture to the meat making its handling easy;
- cleanses the fish meat, reduces the bacterial load;
- inhibits enzymatic reactions and maintains nutritive value by retarding browning reactions;
- sets the natural colour of the product;
- expels the respiratory gases from the tissues, thus helping to improve the vacuum in the can and
- removes the 'raw' flavour of fish.

Blanching in hot water/brine results in appreciable loss of soluble materials. This can be minimised by cooking in steam. Blanching influences the drained weight of the pack. If the fish is not sufficiently blanched, water will be released from the flesh during heat processing which will result in under-weight of the pack, cloudiness of brine, formation of water-oil emulsion etc. Over-blanching, particularly in the case of hot blanched products like shrimp, leads to excessive drained weight due to absorption of water from the filling brine during heat processing. The can may appear swollen and the texture and flavour will be adversely affected.

In general practice, fish like sardine and mackerel are cold blanched in brine followed by pre-cooking in steam. Big fish like tuna are pre-cooked in steam under pressure. Solid salt is added to the fish after filling in the cans. Shrimp and most other shellfishes are blanched in boiling brine.

The hot blanched material, after draining off the blanching medium, is generally cooled under an air blast to drive away the hot water vapour from the surface and thus make the surface dry. This is a very important requirement, as it will influence the drained weight of the product. If the vapour is allowed to condense on the surface it will form a part of the weight filled in the can. Pre-cooked fish like tuna are generally allowed to cool overnight in a cold room. This will make the meat firm enough to render its further handling like removal of skin, polishing the loins, cutting to the size of the can etc. convenient.

9.2.3 Can Filling

The blanched material is filled in clean cans. Weight filled will depend on the specific requirements with respect to the size of the can. It is then covered with a liquid medium like hot brine, oil or sauce. The liquid medium is a constituent of the product and helps in improving its taste, texture and flavour. However, a still greater role it plays is in facilitating proper heat penetration in the product during heat processing. Other additives like flavouring agents, vegetables, etc. may be added with a view to improving the flavour, presentation etc.

Double refined deodorised vegetable oil is the principal filling medium used in fish cans. The oil should not undergo any change during subsequent heat processing and also should not impart any colour or undesirable flavour to the product. Tomato sauce is an important additive in canned sardine, mackerel, oyster etc. It is important that the colour of the sauce does not change during heat processing. The sauce should have a solid content around 30 per cent. The other additives generally used are carboxymethylcellulose, monosodium glutamate, spices, sugar etc. added in specific cases to yield canned products of specified characteristics.

Great care has to be taken in filling the cans. During subsequent heat processing, internal pressure in the can is generated by

- expansion of can contents;
- increase in water vapour pressure and
- expansion of air and other gases in the fish.

This internal pressure is partially balanced by the expansion of the can and the outward movement of the can ends. However, a completely filled can would be subjected to excessive strain. To eliminate this filling is done leaving sufficient headspace over the material.

9.2.3.1 *Precautions in Filling the Cans*

The cans are filled such that a uniform headspace of 6-9 mm is available above the contents. Can ends may bulge if the headspace is too little. This may even cause uneven sterilisation in the product. Too high a headspace also causes problems because too much air in the can will accelerate product deterioration and container corrosion besides adversely affecting the vacuum. The ratio of solid to liquid should be uniform, because this is an important factor influencing heat penetration. There should be no air pocket occluded in the pack that cannot be expelled later by exhausting.

9.2.4 Clinching

While seaming the cans using high speed machines there

is a possibility of the contents spilling over. To prevent this, the can end is clinched to the can body. Clinching involves the can end being partially secured to the body by a single seam on the opposite sides keeping the lid sufficiently loose to allow escape of air and water vapour during exhausting.

9.2.5 Exhausting

Exhausting is the step by which air from the headspace and the contents is removed prior to seaming the can. Exhausting is an essential step because it

- minimises the strain on the can and seams due to expansion of air during heat processing.
- removes oxygen which can, otherwise, accelerate internal corrosion of can and oxidation of fat and vitamins.
- creates partial vacuum in the can. When stored at higher than normal temperature or at higher altitudes the ends of the cans that are not exhausted will expand and present a blown appearance. During exhausting, the air in the can is replaced by steam. On cooling the steam condenses creating a vacuum. Vacuum will ensure that the can ends remain concave or flat even during storage at higher than normal temperatures and at higher altitudes.
- preserves vitamin C (though this is not of any significance in fish)

In the absence of vacuum an apparently blown can will result under these circumstances. Vacuum will also act as reservoir for any gas produced in case of internal corrosion of the can.

Different methods of exhausting are given below.

9.2.5.1 Heat Exhaust

In heat exhausting, the contents of the can are heated at the required temperature before sealing. Heat exhausting can be carried out in two ways. The material to be packed may be heated immediately before filling into the can and then quickly sealed or the can filled with the material is passed through live steam in an exhaust box. The contents get heated, air in the product is expelled and the headspace is filled with steam. The cans are immediately seamed as they reach the exit side of the exhaust box. The heating time is so adjusted that the can contents attain a core temperature of 80-85°C. The cans on cooling will develop a vacuum due to the condensation of the steam in the headspace. In heat exhausting process the cans may be exhausted with clinched ends.

9.2.5.2 *Mechanical Exhaust*

This is most commonly employed where a vacuum sealing machine is used for seaming. The cold material is filled in the can and is subjected to vacuum during the seaming operation.

9.2.5.3 *Steam Injection*

This involves injecting a blast of moist steam into the headspace which blows the air away as the lid is placed in position. Steam is injected through ports around the seaming head of the closing machine. The air from the headspace is swept away and a vacuum develops when the steam condenses in the can.

Heat exhaust is considered the best method of exhausting as the can reaches the retort hot which has a positive influence on subsequent heat processing. Further, the method is simple, less expensive and reliable for obtaining required level of vacuum in the can.

9.2.6 Can Coding

It is a statutory requirement to stamp the can ends with a code denoting the contents, date of manufacture and other details as demanded under the food laws. This can help in identifying the batch and impound such cans if any problem arises during storage. Coding is done in figures and letters using an embossing machine. Coding should not damage the lacquer or the tin coating which will lead to corrosion of the can.

9.2.7 Can Seaming

The object of seaming is to get an air-tight seal between the cover and the body of the container so that microorganisms cannot gain entry into the can. Cans are seamed immediately after exhausting or along with exhausting as in vacuum seaming. Any delay may cause cooling of the can contents. This may lead to air occupying the space giving no room for development of vacuum. A good quality tinplate, adequate sealing compound and an efficient can closing machine combine to produce a strong hermetic double seam. The double seam formed is always examined with reference to standard measurements. The sealed cans are also subjected to the pressure testing for checking the perfection of the seams. Details of formation of the double seam are given in section 9.5.1.5

9.2.8 Can Washing

Cans leaving the seaming machine may have pieces of fish, sauce or oil adhering to the surface. These can contaminate the retort, clog it and become sources of contamination of subsequent batches of cans. Fish pieces sticking to the can surface, when removed later after retorting, may result in peeling away of the surface lacquer, thus exposing a potent area for corrosion. Surface contamination also may prevent the label properly sticking to the surface. Therefore, the cans are washed before retorting in a hot detergent solution like 1-1.5 per cent sodium phosphate at about 80°C. The washed cans are further rinsed in hot water to remove any detergent residue. Even traces of alkaline residue remaining on the can surface can cause corrosion.

9.2.9 Processing/Sterilisation

The sealed cans are heated for a predetermined time-temperature schedule in saturated steam. Thermal processing should take care of the following aspects :

- consumer safety.
- ensuring non-spoilage under ordinary conditions of storage and distribution.
- proper cooking of the product.
- retention of organoleptic characteristics.

Complete sterility is not the aim of heat processing because such a condition will affect the wholesomeness of the product and even may render the product organoleptically unacceptable. The heat treatment should be such that it is sufficient to kill or inhibit microorganisms causing spoilage without overcooking the product.

Bacteria, spore forming and non-sporers, can be present in canned foods and may not spoil the contents under the normal conditions of storage. This dormancy of the bacteria is favoured by fat that encapsulates and renders them insensitive to heat. So long as the product is free from pathogens and otherwise good it can be considered hygienically acceptable. This has given rise to the concept of 'commercial sterility' which is a condition achieved by the application of heat where the food is rendered free from viable microorganisms having public health significance as well as microorganisms of non-health significance capable of reproducing in food under the normal non-refrigerated conditions of storage and distribution. FAO/WHO Codex Alimentarius Commission defines commercial sterility as a condition achieved by the application of heat, sufficient alone, or in combination with other appropriate treatments, to render the food free from micro-organisms capable of growing in the food at normal, non-refrigerated conditions at which the food is likely to be held during distribution and storage. This implies that the recommended process does not kill all microorganisms; some spores of thermophilic organisms may remain and hence the food is not bacteriologically sterile.

The important consideration in canned fish is preventing the growth of *Clostridium botulinum,* a food poisoning bacteria that can produce a highly lethal toxin under anaerobic conditions. A sterilisation process which ensures destruction of *C. botulinum* spores will kill most of the organisms which can spoil canned fish under the normal conditions of handling and storage. A reduction in the population of *C. botulinum* by a factor of 10^{12} is considered essential in canned fish. The process can be considered successful if only one in 10,000 processed cans runs the risk of spoilage.

The time-temperature requirements needed to produce a commercially sterile canned food depend mainly on the following :

- the acidity of the food; acid foods requiring less heat processing compared to non-acid foods.
- number and types of microorganisms present and their heat resistance.
- consistency of foods, solid foods and thick liquids requiring longer periods of heat processing.
- storage conditions of the food after heat processing.

The most important among these is acidity. The level of acidity has a pronounced effect on the survival and growth of pathogenic and spoilage organisms. Canned fish is a low acid food with pH 5.3 or higher and *C. botulinum* is assumed to be present in the pack. In order to kill *C. botulinum* spores the centre of a low acid pack must maintain a temperature of 121°C for a minimum of 2.52 minutes, which is the minimum 'botulinum' cook necessary to reduce the probability of spore survival to 1 in 10^{12} cans. Duration of processing actually employed is generally longer than this to ensure safety of the food.

Sterilising action of steam depends largely on its ability to transfer the latent heat to the surface of the cans where it condenses. Therefore, saturated steam is more efficient than dry or superheated steam. It is also important that uniform distribution of steam should be ensured in the retort. Air should be completely eliminated from the retort as air pockets inside the retort may lead to localised under-processing. Therefore, there should be provision for complete venting out of air. Air in the steam reduces the efficiency of the process because

- air reduces the retort temperature unevenly. Air being heavier than steam tends settle below steam.
- a mixture of air and steam at any temperature is not as efficient as saturated steam at the same temperature.
- air in the retort will reduce the penetration of steam.

However, wherever steam-air mixtures are used for heat processing it should be ensured that an intimate mixture of both is maintained in the retort during processing.

Heat processing is carried out in retorts. As a countercheck for the processing temperature, there should be a provision in the retort to read the temperature directly using a thermometer or thermocouple in addition to the pressure gauge.

Different types of retorts are used for heat processing of canned foods. These include still retorts, agitating type retorts, continuous retorts, hydrostatic retorts etc. The type of retort used may depend on the product being processed, the quantity involved, speed of operation needed etc. The still retort is the oldest type and is still most commonly in use. Still retorts are of two types - horizontal and vertical. They are batch operated. The main difference between the horizontal and vertical retorts is that the former has its door in the end whereas the latter has its door at the top. Even though these retorts can be used only for batch operation, these are considered the best when the quantity of products handled is relatively small.

After loading the cans into the retort, steam is admitted into it. After flushing out all the air in the retort and after closing the drain and steam exit valves, steam pressure is increased to attain the operating pressure and temperature and the retort is maintained so for the required time. The pressure, and hence the retort temperature, is controlled either automatically or manually. However, sole reliance must not be bestowed on the gauge pressure; there should always be a provision to read the retort temperature directly using a calibrated thermometer or thermocouple.

Designs of horizontal and vertical retorts are shown in Fig. 9.1

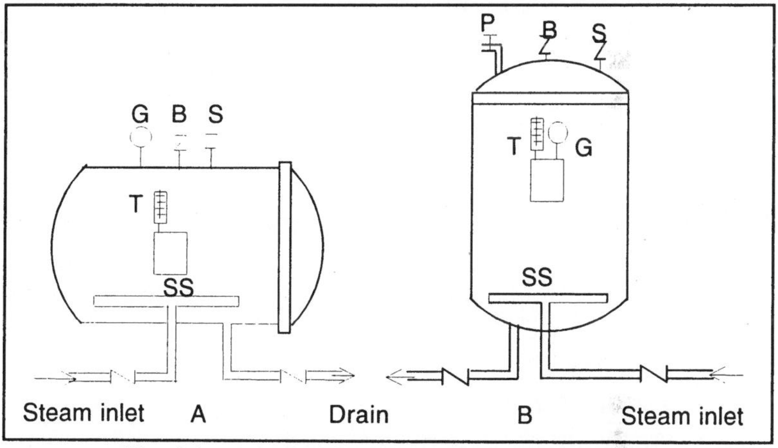

Fig. 9.1 : A horizontal retort and B vertical retort

(SS) steam spreader; (P) pressure release valve; (B) bleeder/vent; (S) safety valve; (T) thermometer; (G) pressure gauge

9.2.9.1 Processing of Non-rigid Containers

With the emergence of non-rigid containers like the flexible pouches as an alternate to tinplate cans appropriate heat processing systems for such containers also became necessary. Overpressure retorts are generally employed to process flexible pouches where the pouches are held in water heated with steam and provided with an overpressure of air to prevent extensive distention of the pouches. The other method employed in the cases of non-rigid containers is processing in air-steam retorts. Air being twice denser than steam will tend to stratify below steam. Therefore a mixture of air and steam maintained at a uniform composition, generally by means of forced circulation within the retorts with the help of fans or blowers, is used. Because of the forced circulation the overpressure required to restrain the package can be maintained by the air-steam mixture.

The orientation of the flexible pouches in the retort is a very important factor. Processing the pouches in a vertical position will tend to bulge the bottom resulting in an increase in the cross-sectional area at the base and will significantly affect the heat penetration characteristics. Use of racks can restrict the pouch thickness within the predetermined limits; however, the majority of the pouches are processed in a horizontal position.

9.2.10 Cooling

At the end of heat processing the cans are cooled as rapidly as possible to about 35°C. Even after shutting off steam supply the temperature at the centre of the can will register a slight increase and the product will get overcooked. In order to prevent this overcooking it is essential that the contents are cooled immediately at the end of heat processing. Rapid cooling will also prevent germination of any remaining non-pathogenic thermophilic spores in their optimum temperature range.

During heat processing pressure builds up inside the can due to expansion of the contents, increase in water vapour pressure and expansion of gases in the headspace. This pressure is counter-balanced by the steam pressure in the retort. At the end of processing when the steam supply is cut off the retort pressure falls rapidly. However, the can contents lose the heat slowly and the internal pressure remains high. At this stage the can seams and sealing compound are very soft and the pressure difference can cause severe strains on the seams with the consequent risk of leakage. This may be avoided if an overriding air pressure is maintained till the internal pressure in the can is reduced sufficiently. When the temperature at the can centre is less than 100°C the air pressure may be removed.

Bacteriologically safe water should be used for cooling the cans. Otherwise the cans will run the risk of spoilage and pathogenic bacteria may gain entry into the can through the minute droplets of water sucked in while developing the internal vacuum. Cans should be cooled to an average temperature of 35°C only and not below that so that sufficient heat is retained by the product to evaporate the water on the can surface and make it dry. Any water retained on the can surface may lead to its external rusting.

9.2.11 Labelling

Cans are labeled to identify their contents. Some canners use printed cans. Can code on the labels instead of embossing on can ends is also in use.

9.2.12 Storage and Distribution

Cans are usually stored for short periods, say 1-3 months, before marketing. This will help the contents to mature, ensure equitable distribution of salt and other additives like spices as also stabilise taste and flavour. This also provides a countercheck on the soundness of cans because, any leak in the cans will show up by this time.

The temperature of storage is directly related to the storage life of the canned products. Considering 10°C a highly desirable storage temperature, it will appear that increase of every 10°C will reduce the storage life by half of the previous. Cans should not be stored at freezing temperatures also as this will result in an unsightly product on opening.

The storage should be well ventilated and be free from moisture. Presence of moisture in the atmosphere will lead to its condensation on the can surface and may lead to subsequent corrosion of the can.

Processed cans should not be cased hot. Loss of heat by radiation from the cans is slow and can lead to a situation called 'stack burning' in the can. Spoilage or deterioration in quality or accelerated corrosion caused by retention of heat in stack of cans or cartons for long periods is referred to as stack burning. This may also provide a favourable condition for the growth of any surviving thermophilic spores.

Cans as a rule should be stored under cool, dry conditions. High temperature and condensation should be avoided as these will lead to thermophilic spoilage and rusting.

9.3 THERMAL PROCESSING OF CANNED FOODS

Preservation of foods by canning, as stated before, is based on the principle of destruction or inactivation of bacteria by the action of heat. Death of bacteria occurs as a result of coagulation of proteins and especially inactivation of enzymes required for metabolism. Vegetative cells of bacteria are easily destroyed at 100°C, but the spores are more heat resistant and require longer exposure to higher temperatures to inactivate them. Generally temperatures above 60°C is considered lethal to vegetative cells and above 100°C for spores.

9.3.1 Acidity Classification of Foods

In deciding the thermal processing requirements of canned foods an important factor to be considered is the acidity of the foods concerned because pH has a profound influence on the heat treatment required to process a microbiologically safe food. Based on acidity, canned foods are classified into four groups.

- *Group 1.* Low acid foods (pH 5.3 and above). The principal commodities grouped under this are meat products, fish products, milk and certain vegetables.

- *Group 2.* Medium acid foods (pH 4.5 to 5.3). Meat and vegetable mixtures, soups, sauces etc. are important in this group.
- *Group 3.* Acid foods (pH 3.7 to 4.5). Canned foods like tomato, pineapple, figs etc. come under this group.
- *Group 4.* High acid foods (pH 3.7 or below). Pickles, grapefruit, citrus juices etc are foods with pH 3.7 and below and hence called high acid foods.

High acid foods like fish marinades and pickles containing acetic, citric or lactic acid do not support the growth of bacteria and hence spore-forming pathogenic microorganisms will not grow in them. Therefore such foods need heat processing only at 100°C or below. Generally the organisms that can grow in such acidic conditions can be destroyed in the temperature range 90-95°C. However, acid tolerant bacteria like *Lactobacillus* and their spores will survive the acidity of acid foods. They will need heating at 100°C for a longer period than usually adopted for acid foods.

Fish pached in tomato sauce is a medium acid food. Such foods will require heat processing sufficient to make them safe against *C. botulinum.*

Canned fish in general, except when packed in medium like tomato sauce, is a low acid food and often will have a pH close to neutral. Therefore canned fish will require heat processing as do low acid foods to ensure safety against *C. botulinum.* All foods likely to have the presence of *C. botulinum* are processed on the assumption that this organism is present in the food and should be eliminated. However, some heat-resistant spore-forming bacteria like *Bacillus stearothermophilus* which are responsible for the flat sour spoilage in canned fish will survive heat processing sufficient to eliminate the spores of *C. botulinum.* Heat treatment sufficient to ensure their complete elimination will result in gross over-cooking of the fish. Therefore, in developing the heat processing requirements for low acid canned foods like fish, the golden rule is to avoid the contamination of the foods with such bacteria by resorting to hygienic handling practices. Only safety against *C. botulinum* is ensured in such cases. This is the principle based on which thermal processing requirements of canned fish are generally arrived at.

9.3.2 Survival of Bacterial Spores During Heat Processing (Decimal Reduction)

Thermal destruction of bacteria or spores takes place in a definite pattern. If a suspension of bacterial spores is exposed to a constant lethal temperature and samples are taken at intervals and the number of survivors are counted it can be seen that its number decreases in a definite pattern with increasing time of exposure, equal percentages of surviving cells, 90 per cent, dying in each successive unit of time. If the logarithms of the number of the surviving spores are plotted against time on a linear scale (or the number of survivors in a log scale against time in linear scale) a straight line graph will be obtained. This is variously called as 'Thermal Death Rate' curve, 'Decimal Reduction Time' curve or 'Survival' curve. The slope of the curve represents the Decimal Reduction Time (*D*) or death rate. *D* is equal to the time in minutes required to reduce the number of survivors to one-tenth of the original at a specified temperature. This is the same as the time required for the curve to traverse one log cycle. *D* value is generally denoted with a subscript of the temperature at which the determination is carried out. For example, D_{121} indicates the Decimal reduction time determined at 121°C. Decimal death rate curve is shown in Fig. 9.2.

From the Decimal reduction time curve it becomes apparent that higher the initial load of bacteria, the longer will be the time required for their destruction. This concept is of great technological and commercial significance since it implies that a standard heat treatment given to two different samples of the same product processed under identical

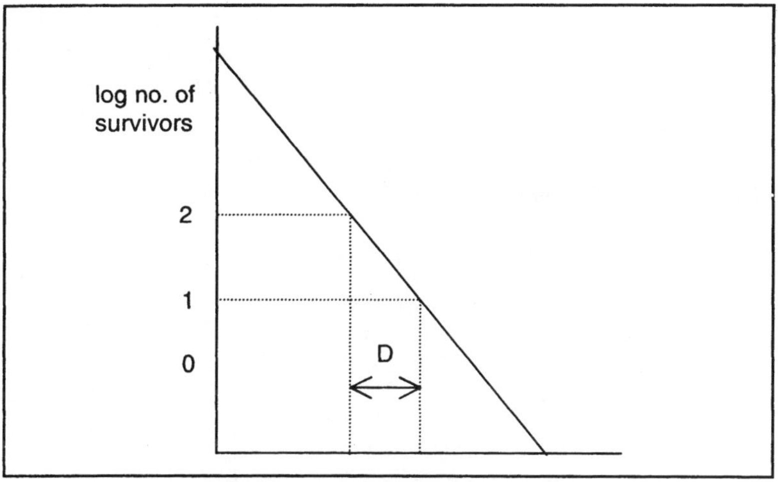

Fig. 9.2 : Decimal death rate curve

conditions cannot have the same effect on the bacterial population if the initial population in the two samples are different. It is therefore necessary to ensure that bacterial contamination should be minimised at every stage prior to heat processing.

9.3.3 Thermal Death Time and the Methods of Estimation of Heat Resistance

Thermal death time (TDT) can be defined as the time in minutes required to inactivate an arbitrarily chosen number of spores of a given bacteria at a specified temperature. Death is defined as the failure of bacteria or spores to reproduce when favourable conditions for reproduction are provided.

9.3.3.1 Methods of Determination of Heat Resistance/Thermal Death Time

Since data on heat resistance of bacteria are essential to determine the processing needs of foods, accurately standardised techniques are required to secure reliable data. Most of the methods employed subject suspension of a known number of spores of the reference bacteria in sealed containers to a constant temperature for varying periods and determine their survival by sub-culturing. If the number of survivors at any stage is to be enumerated, a quantitative count will be taken using a suitable culture method. The composition of the recovery medium and the temperature of incubation are very crucial. The spores or cells exposed to high temperatures will be much more exacting in their nutritional needs than before heating. Some of the important methods employed are described in brief.

• METHOD OF BIGELOW AND ESTY

This is a 'single tube method'. Known number of bacteria or spores in suspension contained in sealed glass tubes are immersed in an oil bath maintained at a constant temperature. A single tube is removed at known short intervals, rapidly cooled and cultured in a suitable medium. The death time is taken as that lying between the time of the last tube showing growth on sub-culture and the first tube showing no growth. A major disadvantage is the problem of "skips" where a tube may give no growth on sub-culture although heated for a shorter time than the tubes showing growth.

• METHOD OF ESTY AND WILLIAMS

This is called the "multiple tube method" and is a modification of the single tube method. Instead of a single tube removed at short intervals in the single tube method, a large number of tubes, around 25-30, are removed at each of the four widely spaced intervals. The percentage of the tubes giving growth for each heating period is plotted on a semi-log paper. The thermal death time of the organism at the temperature employed can be obtained from the resultant straight line graph by extrapolation. The multiple tube method is the one which has been most frequently used in process determinations.

A number of other methods are available to determine the thermal death time of bacteria or spores; however, all of them are based on the same principle, i.e., exposing the bacteria or spores in a suspension to known temperatures for known intervals and sub-culturing them. It is only the mode of exposure, the container and the media etc. that differ. However, a special mention has to be made of a 'Bulbed capillary tube method' in which the spore suspension is sealed into a bulb blown from capillary tube and is placed at the slowest heating point in a can of food. Actual time of exposure of the bulb to the processing temperature is compounded taking into consideration the come-up and cooling times.

The thermal resistance data obtained by some of the above methods can be used to arrive at the *D* value. One method is to directly read from the survival curve (see Section 9.3.2). The other is the calculation method where, considering that the order of destruction is logarithmic, *D* value is calculated from the equation

$$D = \frac{t}{\log a - \log b}$$

where '*t*' is the time of heating, '*a*', the initial number of spores and '*b*', the number of survivors.

9.3.4 Thermal Death Time Data

The heat resistance data of the bacterial spores are important to determine the process variables and also to evaluate the process efficiency. Thermal destruction data for several lethal temperatures are needed for process evaluation. Thermal death times at a number of different temperatures are determined for the same proportion of bacteria employing one of the methods described above. Thermal death times are plotted on a log scale against corresponding temperature in linear scale. Alternately, logarithms of the death times can be plotted against corresponding temperature, both in linear scale. The result in either case is a straight line graph called the 'Thermal death time curve' (TDT curve). Thermal death time for any temperature not listed in the graph also can be obtained from the curve or an extrapolation of the same. A typical TDT curve is presented in Fig. 9.3.

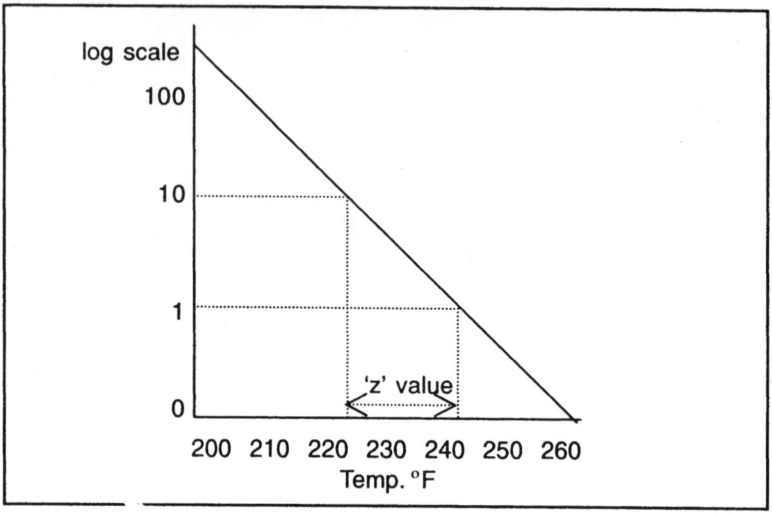

Fig. 9.3 : Thermal death time curve

9.3.4.1 z Value

The slope of the TDT curve is defined as the 'z' value. From the curve it can be seen that *z* is equal to the number of degrees on the temperature scale when the curve traverses one log cycle. In other words, *z* is the change in temperature necessary to cause a ten-fold change in the *D* value. Since an increase in the heating temperature causes a decrease in the *D* value, the *z* value will be equal to the increase in temperature necessary to effect a ten-fold or decimal decrease in the D value. The value of *z* for *C. botulinum* is 10°C by which it is meant that for every 10°C change in the temperature there is a ten-fold change in its death rate.

9.3.4.2 F Value

F value, originally called the sterilising value, is used to denote the time in minutes required to kill an organism in a specific medium at 121.1°C, i.e., its thermal death time. When the z value of the process is 10°C, *F* is denoted as F_o. F_o is equal to the integrated lethal value of heat received by all points in a container.

The total sterilising effect of a thermal process can be considered as the sum of all sterilising effects achieved by all the time-temperature combinations throughout the entire thermal process at all points inside the can. By convention, sterilisation effect is expressed in standard units at 121.1°C so that the entire processing cycle is expressed as being equivalent to holding the product at 121.1°C for a given time. The unit of sterilisation is F_o, which can be defined as the integrated lethal value of heat received by all points inside a container during processing. F_o value of 1 is equivalent to holding the product at 121.1°C for one minute assuming instantaneous heating and cooling.

9.3.4.3 '12 D' Concept

'12 *D'* concept is a method of expressing process lethality requirements. To determine the extent to which the canned foods must be processed, the probability of survival of bacterial spores is taken into account. The bacteria of significance in low acid foods which cover most of canned fish products, as well as medium acid foods, is *C. botulinum* because of its ability to produce a deadly lethal toxin in the food. Therefore, the modern canning practice demands a reduction in the *C. botulinum* spores by a factor of 10^{12} in such foods. This means that the probability of survival of spores of C *.botulinum* must be reduced to one can in 10^{12} i.e., one in a million million cans.

D value of *C. botulinum* spores as measured from the survival curve obtained on heating at 121.1°C is 0.21. Considering an initial loading of one spore of *C. botulinum* per can and that this has to be reduced by a factor of 10^{12}, the heating time required at 121.1°C is $0.21 \times 12 = 2.52$ min

Some of the earlier experiments where the largest possible concentration of spores of *C. botulinum* which the workers could manage to get, and considered equivalent to 10^{12} spores per ml, it had been indicated that it took 2.72 minutes at 121.1°C to reduce the survivors to 10^0, i.e., one. However, with still greater controls over the experiments it could be proved that the time taken was actually only 2.52 minutes, which is equal to 12 *D* of *C. botulinum*.

Now, considering the equation

$$D = \frac{t}{\log a - \log b}$$

$$\text{Or, } t = D\,(\log a - \log b)$$

and that this equation is applied to cans containing one spore of *C. botulinum* per can, it can be seen that the value of '*b*' is reduced to less than zero, but is never zero. A fractional value of '*b*' indicates the probability of survival of the organism in the sample. Assuming that a can contains only one spore of *C. botulinum* and that is subjected to a time-temperature combination equivalent to 2.52 which is 12 *D* of the organism, and substituting the value of 0.21 for *D* in the equation,

$$\begin{aligned} 2.52 &= 0.21\,(\log a - \log b) \\ &= 0.21\,(\log 1 - \log b) \\ &= 0.21\,(0 - \log b) \\ &= -0.21 \log b \end{aligned}$$

$$\text{Log } b = \frac{2.52}{-0.21} = -12$$

$$b = 10^{-12}$$

which means that one spore is reduced to 10^{-12}.

2.52 is the theoretical F_0 value for the spores of *C. botulinum* and this indicates that with this Fo one spore is reduced to 10^{-12} or that 10^{12} spores will be reduced to one. Therefore, it follows that a minimum heat-process equivalent to 12 *D* is sufficient to reduce the probability of survival of the significant bacteria by a factor of 10^{12}; or in other words, a '12 *D*' process is sufficient in canned foods with respect to the reference bacterial spore.

It can also be seen that 2.52 is equal to 12 times *D* of *C. botulinum*. 2.52 is the minimum 'botulinum cook' that should be ensured in low and medium acid foods.

9.3.5 Determination of Thermal Process

All points inside a can do not get heated at the same rate at any given time. The point of greatest temperature lag, that is the slowest heating point in the container, is called the 'cold point'. Since commercial sterility should be ensured throughout the product in the container, the process evaluation is carried out based on the measurement of the temperature attained by this point and integrating the lethal effects. It is assumed that if the cold point receives a heat process sufficient to attain commercial sterility, then all other points in the product in the can have received an equal or greater thermal process and are commercially sterile.

The time required to vent the retort and bring it to the processing temperature is called the 'come-up-time'. The come-up-time will be different for different retorts and can loads. In evaluating the heat process applied to foods the contributions made during the periods in which the contents are heating up and cooling down also should be considered since temperature above 60°C have some killing effect on the bacteria. Data on the heating and cooling phases of the process are required as both these contribute to the total lethality. For this, heat penetration into the cans during processing has to be measured. Therefore, for the determination of the thermal process heat penetration profile in the filled can as also determination of the cold point are very essential.

9.3.5.1 Determination of Cold Point

Heat penetration inside the can is measured using thermocouples. When two dissimilar metal wires are joined at the ends and the ends are held at two different temperatures, a current is set up depending on the difference of temperature between the two junctions. One end can be introduced into the can and the other end will be held at a constant temperature, usually melting ice. Temperature is measured on a potentiometer calibrated to give direct readings of temperature.

To determine the cold point in the can, thermocouples are fixed in such a way that a number of cans are available with thermocouples positioned at every half through three-fourth to one inch from the bottom. Cans filled with food are retorted and the temperature is recorded every minute. Experimental studies using the thermocouples have shown that in products heating by conduction, i.e., in solid packs, the cold point is approximately the geometric centre of the can. In products heating by convection the cold point is on the vertical axis near the bottom of the container.

Profile of heat penetration in cans is presented in Fig. 9.4.

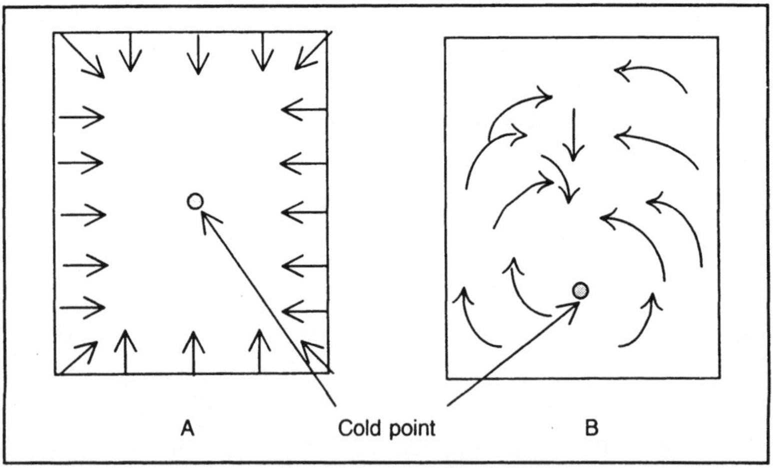

Fig. 9.4 : Heat penetration profile

(A) Conduction heating pack (B) Convention heating pack

9.3.5.2 *Factors Affecting Heat Penetration*

In order that the results obtained with respect to heat penetration pattern in test cans are reproduced in commercial operations an understanding about the factors affecting heat transfer in canned foods is essential. Some of the important factors are discussed.

• FILLING

Heat penetration in a can is affected by the temperature gradient between the can and the retort, the rate becoming slower as the temperature difference decreases. Increase in the retort temperature results in more rapid heat penetration. Due to the effect of the temperature gradient on the rate of heat penetration, a cold filled can will reach the retort temperature in about the same time as a partially heated can. However, the importance of hot filling the cans cannot be undermined, since hot cans will reach a lethal temperature more rapidly than cold filled cans.

Another factor affecting heat penetration is the ratio of the solid to liquid and the arrangement of the solids in the liquid. Solids loosely packed in a liquid will heat more rapidly than when it is tightly packed. Overfilling with solid materials will result in under-processing.

• INGREDIENT RELATED FACTORS

There are several ingredient-related factors affecting heat penetration. Fatty tissues are poor conductors of heat compared to lean tissues. Substance which undergo a change of state during heating, e.g., sol to gel, will adversely affect the rate of heat penetration. Solids absorbing water changing the solid/liquid ratio in the cans is another factor affecting the heat penetration.

• PREPARATION RELATED FACTORS

The use of partially thawed frozen material will cause a considerable temperature lag. Differences in handling practices over those tested in laboratory or pilot scales also may be a factor causing changes in the heat penetration pattern.

• PROCESSING RELATED FACTORS

The 'come-up' time need not be the same in commercial production schedules as with the test cans. It has, more or less, been established that 42 per cent of the come-up time should be added to the process time at the retort temperature.

Agitation of the cans during heating significantly increases the rate of heat penetration in liquid packs. Agitation helps to transfer the heated portions of the contents to the cooler regions of the can. A very efficient method of increasing the rate of heat penetration is end-over-end agitation. The cans are arranged with their longitudinal axis aligned as the spokes in a wheel so that they rotate end-over-end as they move around the circumference.

Other factors of great significance in heat penetration are the shape and size of the containers as the rate of heat penetration is affected by the ratio of the surface area of the container to its volume. Smaller cans heat more rapidly because of the large surface area in relation to the volume of the can.

9.3.5.3 Process Evaluation

The time theoretically required for destroying any specific bacterial spore present in the food in the can at any given temperature can be calculated making use of the lethal rate of the temperature, TDT curve and, the heating and cooling curves.

The rate of destruction of an organism per minute at any given temperature is the reciprocal of the time in minutes required to destroy it at that temperature. The sterilising effect of the thermal process $F_{0,}$ can, therefore, be computed by integrating the combined lethal effects of exposure at all time-temperature combinations throughout the process. It has been stated earlier that the spores of *C. botulinum*, the most heat-resistant toxin producing reference organism in canned fish, needs heat processing equivalent to F_0 of 2.52 for their destruction. F_0 of 2.52 can be attained by heating the contents to 121.1°C instantaneously, holding at that temperature for 2.52 minutes and then cooling it instantaneously. The required F_0 can be achieved by any other process combination which can give F_0 equivalent to 2.52. From the knowledge that F_0 at 121.1°C is unity and that for every 10°C temperature change there is a ten-fold change in the death rate of the organism, the F_0 equivalent of other temperature-time combinations can be worked out. If the processing is done at 111.1°C instead of at 121.1°C (i.e., 10°C below 121.1°C), the product will need heating for 10 times as long as at 121.1°C to get the same killing effect, i.e., $10 \times 1.0 = 10$ minutes. Alternately, if heated at 131.1°C (i.e., 10°C above 121.1°C), the heating need will be only one-tenth of that at 121.1°C, i.e., only 1/10 minutes. In real terms a F_0 of 10 in solid pack tuna can be attained by processing for 74 minutes at 116°C, or 50 minutes at 121.1°C. The sterility effect in both cases is the same and is equivalent to holding at 121.1°C for 10 minutes under conditions of instantaneous heating and cooling.

The main procedure employed in estimating the lethality of the heat process involves measurement of the temperature at the point of greatest temperature lag inside the can and integration of lethal effect at this point by graphical or mathematical procedures. Heat penetration data during the heating and cooling cycles are collected for this purpose. Different methods are employed to arrive at the process lethality such as

- the classical calculation method.
- the General method or the Graphical method.
- mathematical method/Formula method.
- nomogram method.

• CLASSICAL CALCULATION METHOD

This is a very simple procedure. The data on the temperature at the point of latest temperature lag is collected for this purpose. Any temperature reached at this point may have a lethal value and hence the procedure involves integrating the lethal effect of the temperature attained by this point.

• THE GENERAL METHOD/GRAPHICAL METHOD

In the Graphical method of process evaluation the procedure employed is collecting the heat processing data and the conversion of the product temperature into lethal rates. The temperature profile in the product is obtained by recording the temperature at the cold point including the 'come-up-time' which is the time elapsed from placing the cans into the retort and starting to heat, until the retort reaches the processing temperature. The temperature at the cold point is recorded initially and then at frequent intervals, depending on the rate at which the product is heated. It is not possible to heat the cold point to the exact retort temperature and hence the cans are processed until the temperature at the cold point is within 1°C of the retort temperature. At this stage steam is shut off and cooling is started. The temperature is recorded during the cooling phase as well. It is important that the cooling procedure employed should be the same as practiced in the actual commercial operations.

From the TDT curve of the reference organisms, its thermal death time at any temperature can be found out. The lethal rate of the organism will be the reciprocal of the number of minutes required to kill the organism at that temperature. Alternately the lethal rate for any temperature can be obtained from lethal rate Tables available.

A heating-cooling curve can be constructed substituting lethal rates for temperature against time in minutes on an arithmetic graph paper. The graph obtained is called 'lethality curve'. A typical lethality curve is shown in Fig. 9.5. The area under the curve corresponding to a value of one is called a "unit sterility area". The area under the curve can be measured by counting the squares and estimating the factions of whole squares. This value is then converted to the lethality or F_0 by multiplying by the scale dimensions for each square.

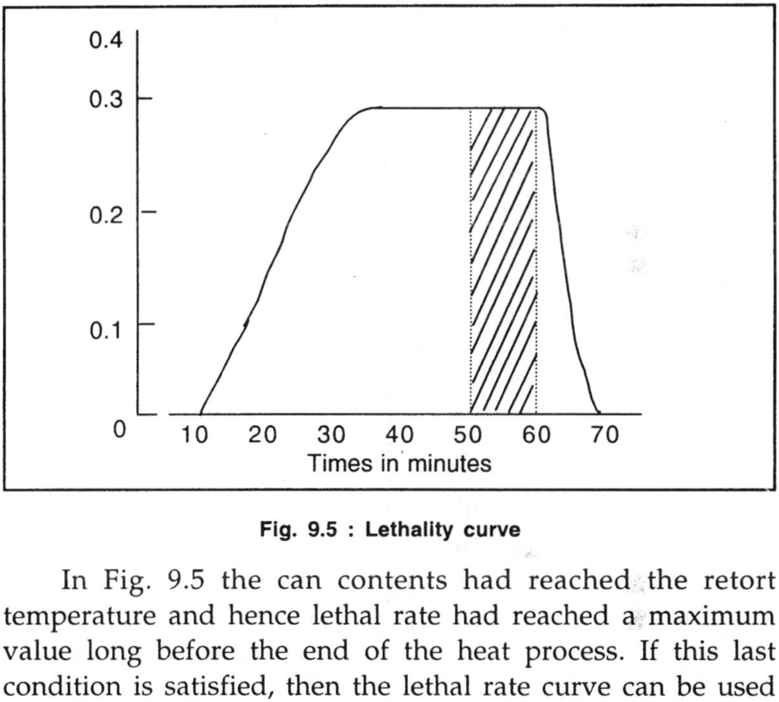

Fig. 9.5 : Lethality curve

In Fig. 9.5 the can contents had reached the retort temperature and hence lethal rate had reached a maximum value long before the end of the heat process. If this last condition is satisfied, then the lethal rate curve can be used to predict the F_0 of a process of say 50 minutes by measuring the area under the curve between 50 and 60 minutes and subtracting this from the total area. Similarly, the F_0 of a 70 minutes process can be found by adding such an area to the total area.

However, the lethality curve cannot be used to predict the F_0 of a shorter or longer process when the can centre temperature does not reach the retort temperature before cooling begins or other changes are made such as different retort temperature or can size.

• FORMULA METHOD

Formula method of determination of processing time involves application of data from the heat penetration curve to an equation and calculating the process achieved mathematically.

$$B = f_h (\log jI - \log g)$$

Where B is the process time in minutes measured from the corrected zero of the process,

f_h = slope of the heat penetration curve,

jI = correction factor obtained by extending the heating curve to intersect the time at which the process begins,

g = value in degrees below the retort temperature when the straight line portion of the heating curve intersects the time at which the process ends.

The slope of the heating curve and the point where the extension of the heating curve intersects the time at which the process begins can be obtained from the heat penetration curve. g is obtained by the difference, in degrees, between the retort temperature and the maximum temperature attained by the food at the cold point in the can.

By solving the above equation substituting appropriate values the thermal process adequate for a product can be calculated.

Nomogram method is considered the most rapid one. The data from TDT and heat penetration are applied to a graphic representation of these matters making a relation.

• MICROBIOLOGICAL METHOD

In order to establish the adequacy of the thermal processing arrived at by any of the several calculation methods available it is desirable that the process so established is confirmed by application on inoculated packs. In cases where heat penetration data are not readily available, microbiological methods will be the most reliable in estimating process lethality. There are two methods used, one in which the spores are added directly to the food and the other in which the spores suspended in a buffer and contained in a small bulb of capillary tube are placed in the food.

• INOCULATED PACK METHOD

An inoculum of the spores of the spoilage organism of importance to the food concerned is placed approximately at the cold point in the container. It is important that the spores should be capable of germination and growth in the food concerned and ideally spoilage should be accompanied by gas production so that the result can be ascertained by the presence of blown cans. An adequate number of cans will be prepared and heat processed so that two sets of cans are available at five minutes interval over the theoretically calculated time-temperature schedule and two sets below that. For example, if the time required at 121.1°C is 40 minutes, cans will be processed at this temperature for 30, 35, 40, 45 and 50 minutes. The cans will be incubated at a temperature optimum for the organism inoculated. After four weeks, the containers not evidencing spoilage will be subcultured. Results can be compared with the theoretically calculated values and their adequacy can be confirmed. Usually a margin of safety is allowed beyond the minimum heat treatment required. However, the process recommended will be successful only for the concentration of the spores used and, often might not take care of any gross contamination.

• THE SPORE BULB METHOD

Bulbed capillaries containing the spores of the heat resistant reference organisms are implanted in solid pieces of food or positioned at the slowest heating point in the food. By comparing the number of surviving spores and the initial number of spores present, the lethality of the process can be evaluated.

9.4 IMPROVEMENTS IN HEAT PROCESSING TECHNIQUES

9.4.1 High Temperature Short Time (HTST) Processing

The relatively long time at temperatures upto 121°C employed to destroy the spoilage organisms in the conventional heat processing may affect the quality of food in appearance, flavour and nutritive value. One of the major achievements in the efforts to overcome this problem is the development of a High temperature short time (HTST) processing technique. The basic approach in this technique is to achieve more rapid heating of foods in order that higher processing temperatures and reduced time of heating can be employed. For every increase of 10°C in temperature beyond the maximum conditions of growth, there occurs a 10-fold increase in the destruction of microorganisms while resulting only in doubling the rate of chemical reactions responsible for product deterioration. Therefore, heat treatment at higher temperatures for shorter periods of exposure should effect greater retention of the natural characteristics of this product than a process of equal lethality attained by heating at a lower temperature for longer periods.

Two methods are employed in the HTST processing techniques.

In one method, the food is heated to high temperatures in bulk or on-stream. It may, or may not, be partially cooled before filling. The residual heat in the product should be sufficient to sterilise the container and the lid. The hot product is filled and sealed in the can. For products with pH above 4.5 filling temperature is maintained above 100°C and the operation is carried out under pressure to prevent the product from boiling.

In the second method the food which is previously sterilised by heating to high temperatures in bulk or on-stream is cooled, filled and sealed in sterile containers in a microbe-free atmosphere. The process is called aseptic canning. The process involves heating the product by pumping it through a heat exchanger so that its temperature is raised to a controlled and predetermined level, passing it directly through a holding section to provide a predetermined residence time to effect sterilisation and cooling it to at least 35°C. The containers and covers are sterilised with superheated steam and the relatively cool sterile product is aseptically filled into the sterile container, and sealed in an atmosphere of saturated or superheated steam.

The process is advantageous to heavy or viscous products which are adversely affected by sterilisation in sealed containers.

9.4.2 Use of Agitating Retorts

There is a limit up to which the temperature can be increased in still retorts. Higher temperature means maintenance of higher steam pressure in the retort with the consequent risks involved. Even otherwise, if the temperature is raised beyond a limit in still retorts, it will result in overheating the contents near the can walls producing the so-called 'burnt' flavours. The probability of such overheating while using high temperature can be overcome if an agitating retort is used in place of still retort. Agitation increases the rate of heat transfer from the container walls to the product by continuously renewing the surface in contact with the container.

9.4.3 Use of Hydrostatic Continuous Cooker-Cooler

In the hydrostatic cooker-cooler the cans are conveyed by a single chain conveyor provided with flights for carrying the cans through a 'W'-shaped pressure vessel at a rate corresponding to the required processing time. Steam pressure necessary to sterilise at the required temperature is balanced by a hydrostatic or water head pressure in the heating and cooling water legs.

Steam is admitted to the central chamber from top with sufficient pressure to maintain the water head at the required height so as to maintain the recommended steam pressure and temperature inside the chamber. The speed of the conveyor carrying the cans is so adjusted that the cans remain in the steam chamber for the required length of time. Cooling water is admitted and made to move in the opposite direction of the can movement. The water leaving the system will be hot and this hot water can be used for initial heating of the cans, thus ensuring optimum use of heat energy. An outline of the hydrostatic cooker is presented in Fig. 9.6.

Hydrostatic cookers are designed to accommodate different can sizes without major equipment change. These are the least complicated type continuous cookers. They require only the minimum floor area for installation. However, hydrostatic cookers need a lot of mechanical accessories like feeder, conveyor as also their control systems. This is considered the major disadvantage with such cookers.

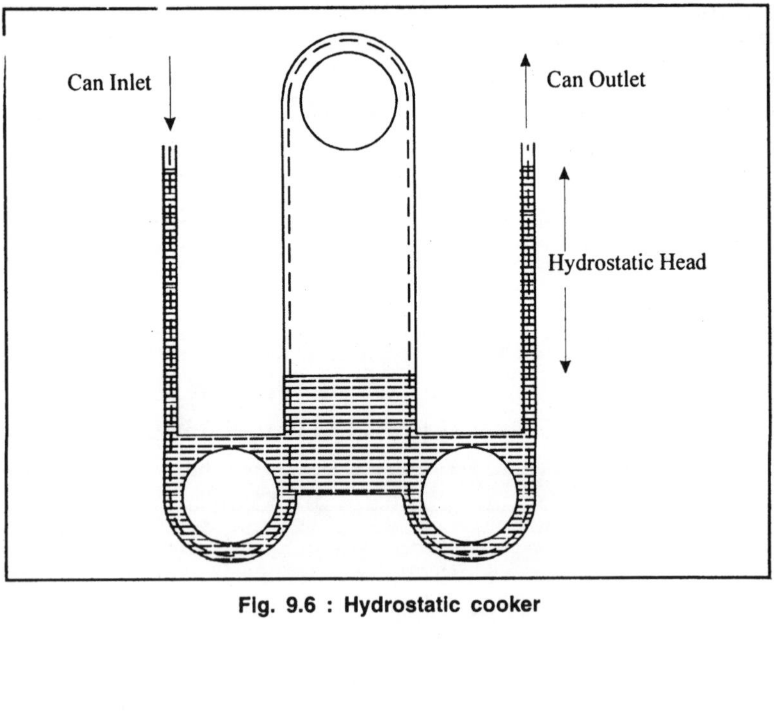

Fig. 9.6 : Hydrostatic cooker

9.5 CONTAINERS FOR CANNED FISH

The metal cans popular in the earlier periods were the hand-made "hole and cap" containers. They had all the joints soldered and the food used to be filled through a small hole in the end. The end is subsequently covered with a cap and is soldered in place. Difficulty in filling large pieces, difficulty in cleaning the cans, possibility of contamination of food with solder and flux etc. were the main disadvantages encountered with this can. Development of "open top sanitary" (OTS) cans having a cylindrical body with soldered lock seam on the side and unsoldered double seam ends made of tinplate is an important development in can making and are now the most popular containers for canned foods. Because of this construction, no flux or solder comes into contact with the food contained by them and are easy to clean, and hence 'sanitary'. However, the term 'tinplate' is a misnomer. In fact cans are made from a sheet of iron (base plate) coated on either side with tin giving it a final composition of 98 per cent steel and 2 per cent tin. The thickness of steel plate varies from 0.19 mm to 0.3 mm depending on the size and height of the can. An ideal container should have the following features :

- strong enough to protect the contents during handling and transportation.
- light enough for easy and economic handling.
- impervious to air, moisture, dust and microorganisms, once the can is sealed.
- should have a pleasing and sanitary appearance.
- internal lacquer should not impart colour, flavour or toxicity to the contents.
- should withstand the pressure and temperature of heat processing.
- should be inexpensive enough to discard after use.
- should be capable of being hermetically sealed at high speed.

The OTS cans meet most of these requirements and are most popular in the canning industry. In addition to the above the metal cans also lend themselves to high speed mechanised handling, filling, sealing and casing.

Cans can be of three-piece construction or of two-piece. The three-piece cans are generally cylindrical with two lids attached to the ends of the cylindrical body by double seaming. Two-piece cans are made in different shapes like cylindrical, oval or flat rectangular.

9.5.1 Can Making

9.5.1.1 Base Plate

Corrosion resistance, strength, durability etc. of the tinplate depends on the chemical composition of the steel. The base plate used in can making should be of the can making quality (CMQ) steel. The composition of the steel plate should be such that it should conform to the following specifications with respect to the content of other elements :

- carbon 0.04-0.12 per cent
- manganese 0.25-0.6 per cent
- sulphur 0.05 per cent max
- phosphorus 0.02 per cent max
- silicon 0.01 per cent max
- copper 0.08 per cent max

With higher contents of copper and phosphorus, the steel plate becomes more prone to corrosion. Higher contents of phosphorus can, however, impart greater stiffness to steel plates which is desirable in certain applications. Low metalloid content steel with phosphorus content 0.02 per cent called Type MR Quality is used in the manufacture of fish cans. Base plate for can making is made by the cold reduction method. Cold reduced plate has the following advantages over hot reduced plates.

- superior mechanical properties enabling the use of thinner plate to make can without loss of strength.
- more uniform in gauge thickness.
- improved resistance to corrosion.
- less variation in sheet thickness.
- takes on uniform thin tin coating.
- better appearance.

9.5.1.2 Tin Coating

The base plate is coated on either side with tin. The thickness, uniformity and the manner in which the coating is applied affect the resistance of tinplate cans to perforation and corrosion. Dipping in molten tin and electrolytic deposition are the two methods employed for tin coating. In the hot dipping process the base plate is conducted through a layer of zinc chloride flux into the molten tin. The tinned sheet coming out of the molten tin is passed through a layer of palm oil. As the sheet pass through this stage it is given a 'squeezing' action by tinned steel rolls which largely regulate the final thickness of tin coating.

In the electrolytic process, which is the more popular method of tin coating, the steel plate is first given a light electrolytic pickling in dilute sulphuric or hydrochloric acid and is then tinned by passing through a series of electrolytic cells filled with either dilute acid or alkali. Pure tin is used as the anode. The moving base plate is the cathode. After tin coating a heat treatment is given to the plate to give a momentary fusion leading to the formation of an intermetallic tin-iron compound, the character of which influences the corrosion resistance of the tinplate. This will also make the tin surface bright. To protect the tin surface from oxidation a chemical passivation treatment is applied by spraying or dipping in a chromic acid or sodium chromate solution. After rinsing the surface and drying, it is given a very thin coating of oil.

The requirement of tin coating thickness can be the same or different for the internal and external surfaces of the can. When the thickness is the same on either side it is called 'even coating' and when different, 'differential coating'. Differentially coated cans are in most common use. In differential coating the food contact surface is given the higher coating of tin. The weight of tin deposited is expressed as grams per square metre (GSM). A recommended differential coating of tin for fish cans is *D* 11.2/5.6 which indicates a coating thickness of 11.2 GSM tin on the food contact surface and 5.6 GSM on the outer surface. When the coating thickness is the same on either side it is indicated as *E*; e.g., *E* 7.5 indicates a coating of 7.5 GSM tin on either side of the base plate. Profile of a an electrolytic tinplate is shown in Fig. 9.7.

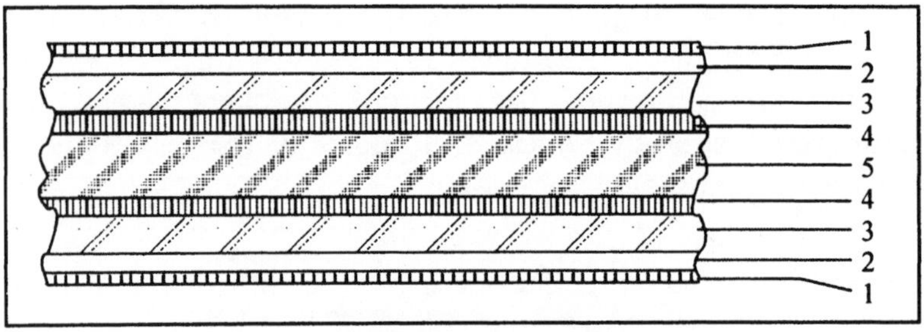

(1) oil film; (2) passivating film; (3) free tin layer; (4) tin-iron alloy; (5) steel base

Fig. 9.7 : Profile of electrolytic tinplate

9.5.1.3 *Lacquering*

Certain foods react with the metal of the can resulting in corrosion of the metal ultimately leading to its perforation. Therefore, it is customary to coat the inside of the container with a thin film of an inert material to prevent the contact of the food with the metal. This is done by lacquering the tinplate before it is cut and made into cans.

Two types of lacquers are used in the food cans, the acid-resistant (AR) lacquer and the sulphur-resistant (SR) lacquer. The former is intended for packing acid foods like fruits. Foods like fish or meat release low molecular weight sulphur-bearing compounds during processing which will react with exposed tin or iron producing their sulphides which are coloured compounds. In cases of extreme build-up, the coloured compound may even get mixed with the product making it unappealing. To prevent the formation of this discolouration fish cans are coated inside with a sulphur-resistant (SR) lacquer.

Many types of SR lacquers are available; however, the most common used in fish cans are the oleoresinous C-enamels. C-enamels contain zinc oxide which will react with any sulphur compound liberated during heat processing producing zinc sulphide which is white in colour. The colour of the product will remain unaffected.

The lacquer is applied on the tinplate by roller-coating. The lacquer-coated plates are baked at 180°C for a predetermined period for setting the lacquer firmly on the plate. The AR or SR lacquer, as the case may be, is applied only on the food contact side of the can. The exterior of the can is usually coated with a rust-resistant decorative lacquer. The lacquer applied on the food contact surface should not react with the contents and should not impart any taste, flavour or colour to the product, nor should it produce any toxin. It should also be resilient enough to stand the rigours of can forming, embossing on the can end etc. The decorative lacquer on the external side should be able to take on printing, lithographing etc.

9.5.1.4 'Three-piece' Cans

Most of the operations in the can making process are carried out by automatic machines. Tinplate is first cut to the size of the can. This is called the body blank. The body blanks are cut absolutely square. The body blanks are passed through a flexing machine which helps to remove any irregularity in the plates and work-hardens them slightly.

The corners of the blanks are then notched at each corner on one edge, slits being cut a little away from the ends at the other edge. By doing so, when the lock seam is formed, the ends will comprise of only two thickness of the metal. The blanks are then edged, i.e., a hook is formed on each of the shorter edges by bending the edges in the opposite directions. The hooks are coated with a flux and the blank is formed into a cylindrical body by locking the bent edges tightly together. The locked hook is hammered, a second coat of flux is applied on the outside of the lock seam and is soldered in a solder mill. The excess solder is removed from the seam by a rotating brush immediately after leaving the solder roll. The solder is solidified in a blast of cold air. The locked seam formed will have four thickness of the metal, whereas at the notched end it will have only two thickness of the metal.

Circumferential beads are made on certain types of can bodies, especially bodies of larger cans, to strengthen the body by making it into a column of shorter can bodies. This is a check against handling abuse and paneling pressures.

The cylindrical can body is then flanged by which a flared rim is produced on both ends of the can. The can body is placed between two flanging punches which fit into the ends of the can. The punches are moved slightly towards each other forcing the edges of the can body outwards. The sequences of operations in the can body making process are presented in Fig. 9.8.

• CAN ENDS

Can end is stamped out from tinplate sheet and the edge is curled inwards. The concentric expansion rings on the ends are formed simultaneously while stamping out the ends. The inside of the curl is lined with a sealing compound that is a rubber solution or emulsion. On drying in an oven, this forms a gasket and ensures an air-tight seal in the finished seam.

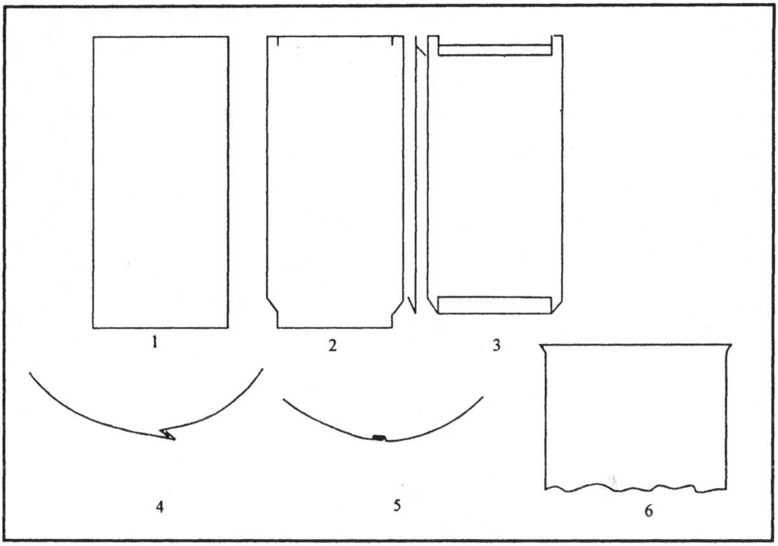

(1) body blank; (2) notching; (3) edging; (4) forming; (5) formed; (6) flanging

Fig. 9.8 : Sequences in can body making

The lining compounds are of two basic types, solvent based and water based. The former is the most widely used because of the volatility of the solvent and resistance of the compound to water. A typical base formula for the lining compound is

Rubber	20-25 per cent by weight
Filler and pigment	50-70 per cent
Resin	10-25 per cent
Plasticiser	<1 per cent
Antioxidants	<1 per cent

9.5.1.5 Double Seaming

A double seam may be defined as that part of the can formed by joining the body and end components, the hooks of which interlock and form a strong mechanical structure. The can ends are attached to the body using a Double Seaming machine.

A double seamer consists of a base plate which supports the can to be seamed, a chuck which fits firmly into the countersink of the can end and a pair of seaming rolls or grooved wheels which operate in succession. Can sealing is carried out in two stages. The base plate raises the can body with the end in position causing the chuck to fit into countersink of the can end. The curl of the can end fits over the flange of the body. The seaming rolls operate in succession. The first seaming roll which has a deeper groove comes into operation and forms mating hooks on the can and its end, i.e., tucks in the curled rim of the can end under the flange of the can body. The second roll which has shallow groove then comes into operation and presses and flatten together the hooked edge of the end and the can body against the chuck forming the final hermetic structure.

The finished double seam consists of five thickness of the metal. However, where it coincides with the body seam, seven thickness will be present. If the blank is not notched, the double seam where it meets the lock seam will consist of eleven thickness of the metal. This may lead to strain on the seam with the consequent risk of leakage and spoilage of the can contents. This underscores the importance of notching in can making. Basic design of a Double Seaming Machine is shown in Fig. 9.9 and the profile of a formed double seam is shown in Fig 9.10.

There are very strict tolerances set down for can seams. A bad seam will result in spoilage and loss of products. No amount of good product preparation and heat processing can compensate for faulty seaming.

9.5.2 Two-piece Cans

Shallow profile cans with a seamless body and one lid on the top are popular for packing fish. The bottom of the can becomes an integral part of the can body and the end is secured on top by double seaming. Advantages of the two-piece cans can be summarised as follows :

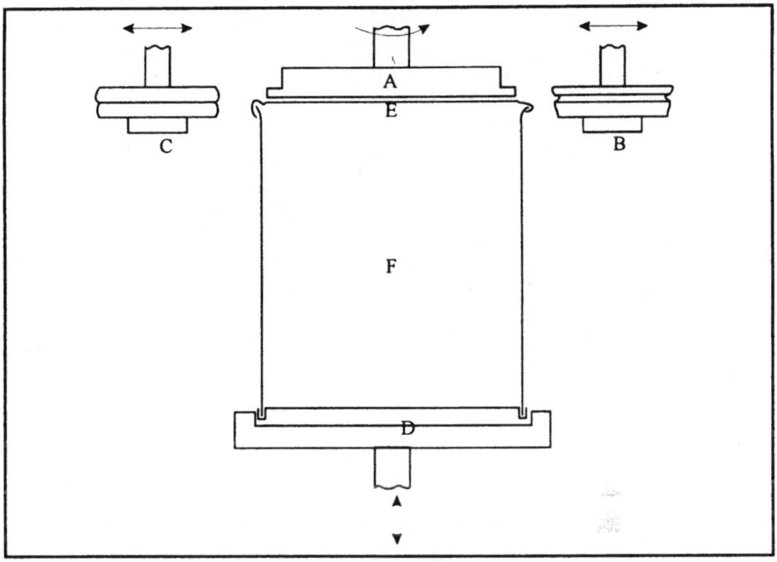

(A) seaming chunk; (B) first operation roll; (C) second operation roll; (D) base plate; (E) can end; (F) can body

Fig. 9.9 : Design of double seaming machine

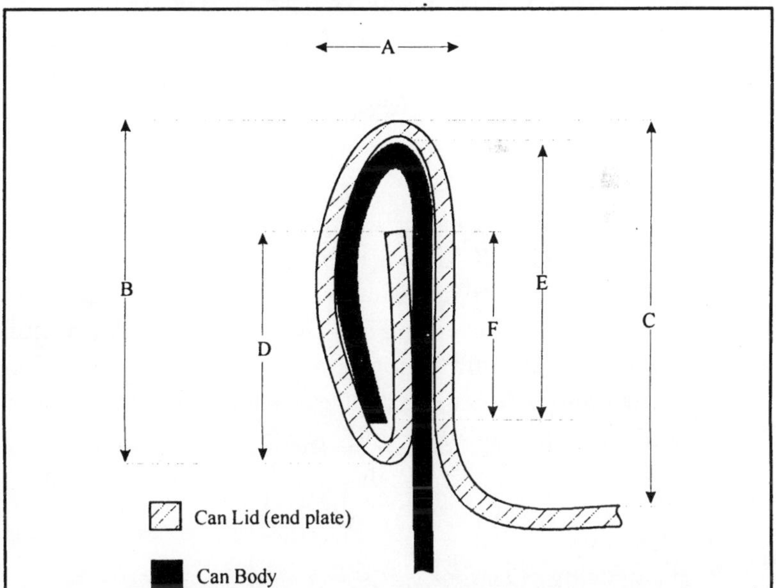

(A) seam thickness; (B) seam length; (C) countersink depth; (D) end hook length; (E) body hook length; (F) overlap

Fig. 9.10 : Seam measurements

- elimination of two seams reduces the possibility of leakage and a potential source of contamination with lead is removed.
- the cans have a better aesthetic appeal presenting a smooth profile and streamline appearance.
- the absence of the side seam permits uninterrupted print decoration on the external surface.
- bottom of the can can be designed and formed for better stackability.
- less metal is used in its construction due to elimination of the overlaps at the two seams.

A major disadvantage with two-piece cans is the difficulty in manufacturing tall cans.

Two-piece cans are 'drawn' cans made by cutting, drawing and flanging in one operation. Because of the strain it is subjected to during the drawing process, the aluminium or tinplate used for can making should be heavier and more ductile than that used for round cans. According to the manufacturing techniques employed, two types of cans can be differentiated.

9.5.2.1 Drawn and Wall Ironed (DWI)

The manufacturing process employed in DWI cans includes the following steps :

- blanking out a disc of metal from the sheet.
- drawing the disc into a cup.
- forcing it by means of a punch through a series of dies, each one being slightly smaller than the preceding one, thus elongating the wall by stretching or ironing action. The process results in reducing the thickness of the plate and increasing the height.
- trimming the body to the required height.
- cleaning to remove the lubricant.
- flanging.
- lacquering. Lacquering is done by spraying.

9.5.2.2 Drawn and Redrawn (DRD)

In the 'drawing and redrawing' process a blank is first drawn into a cup; but the excess metal made available by reducing the circumference of the body is used to elongate the wall, thus maintaining the thickness of the metal. A particular size of can can be made by a series of press operations, the number being determined by the nature of the material and the depth/diameter ratio. Thus, a shallow can can be made by a first draw followed by a redraw, a deeper can by a draw followed by two redraws.

Two-piece cans can be of different shapes - rectangular, oval, oblong etc. These cans require special type of double seamers for sealing the lids on them. In such seamers there will be four seaming rolls, two identical ones each on the opposite sides. The cans will remain stationary with the chuck covered by the end. The seaming rolls will move round the cans, the two opposite pairs in succession. There is a possibility of uneven pressure exerted by the seaming rolls, particularly when the pair of rolls are on the shorter width side of the can. This may cause imperfections in the seam leading to probable leakage through the seams.

9.5.2.3 Easy-Open Ends (EOE)

The easy-open ends made of aluminium which are popular in beverage cans led to the development of such ends for food cans including fish cans. The EOE have circular incision which enables the whole central panel to be removed by pulling an attached ring. However, there is a serious technical limitation in using aluminium EOE with tinplate cans because of the probability of bimetallic corrosion. Tinplate versions of EOE have now been developed; however, their use in thermal processed foods is not yet widespread.

9.5.3 Alternate to Tinplate Cans

Of the several materials proposed as alternates to tinplate in food can making, tin-free steel and aluminium are the most popular ones.

9.5.3.1 Tin-Free Steel (TFS)

Tin-free steel, an important alternate to tinplate, has a steel base with a chromium/chromium oxide coating on the surface replacing the tin in the conventional cans. TFS is made by electroplating cold rolled steel sheet with chromium in chromic acid. The appearance of the sheet is bright or semi-bright as compared to tinplate. Because of the low abrasion resistance of TFS it needs protection by a lacquer film. The surface of TFS provides an excellent substrate for lacquer adhesion which ensures superior performance in terms of products compatibility for many food products. However, TFS cannot be soldered or easily welded without scraping off the chromium coating.

TFS is also made with nickel coating instead of chromium.

9.5.3.2 Aluminium Cans

By far, the most important and most extensively used alternate to tinplate container is the aluminium can. Aluminium used for making food cans is often reinforced by alloying with manganese, magnesium or chromium. Aluminium alloyed with manganese is most commonly used. Alloyed aluminium is first given an anticorrosive treatment, usually anodising in dilute sulphuric acid. The corrosion resistance of aluminium is attributed to the thin layer of oxide imparted by this treatment. To enhance corrosion resistance this is further coated with a lacquer suitable for the food being packed.

Aluminium alloy cans have the following advantages over the tinplate cans:

- light weight; only little more than one-third the weight of a tinplate can of similar dimensions.
- attractive appearance.
- resistance to atmospheric corrosion.
- not stained by sulphur-bearing products.

- non-toxic; does not impart metallic taste or smell to the product.
- can be shaped into containers by several different methods.
- easy to open.
- recyclability and better scrap value.

However, aluminium cans are not free from disadvantages as well, some of which are:

- heavier gauges are required to obtain strength comparable to tinplate.
- not highly resistant to corrosion, some foods require lacquering or some other form of protection.
- aluminium severely bleaches some food products.
- softer than tinplate and hence needs careful handling.
- special precautions are needed during heat processing to avoid any permanent distortions.
- service life is less than that of tinplate cans for most aqueous products.

Different methods are employed to manufacture aluminium cans like the built-up, shallow formed, deep drawn and impact extruded containers. The containers used in fish canning are mostly made by the shallow forming process. Shallow round or rectangular cans are the preferred ones in fish canning.

Container strength is one factor which limits the use of aluminium in can making. Aluminium is best suited for shallow profile drawn cans as those used for packing sardine.

• CAN SIZES

There are several sizes and shapes of cans used for different purposes. Can sizes are generally denoted by two variables, the diameter and the height in that order. The trade names and dimensions of cans commonly used for fish in India are summarised in Table 9.1.

Table 9.1 : Trade Name and Dimensions of Cans

Trade name	Overseam dimensions (mm)
4 ½ oz. Shrimp	77 × 56
8 oz. Shrimp	77 × 60
1 lb. Jam	77 × 90
No.1 Tall	77 × 116
8 oz. Tuna	87 × 43
¼ Dingly	108 × 80 × 22
½ oval	90 × 150 × 81
½ oblong	140 × 57 × 81

9.5.4 Retort Pouch

A relatively recent approach in container development for heat processed foods is flexible thermoplastic films and aluminium foil laminate. A retort pouch can be defined as a container made of 2, 3 or 4-ply material which, when fully sealed, will act as a hermetically sealed container that can be heat processed at similar temperature and pressure as for metal containers. The retort pouch has the advantage of metal cans and plastic boil-in-bag. The material used in retort pouch must also provide superior barrier properties for a long shelf life, seal integrity, toughness and puncture resistance. Typical pouches used have the following configurations :

2-ply 12 μm nylon or polyester/70 μ m polyolefin

3-ply 12 μm polyester/12 μ m aluminium foil/70 μm polyolefin

4-ply 12 μm polyester/12 μ m aluminium foil/12 μm polyester/70 μm polyolefin

The commonest in use is a 3-ply pouch which is a sandwich of thin gauge aluminium between two thermoplastic films. The outer ply generally made of polyester provides barrier properties as also mechanical strength. The aluminium layer protects the material against gas, light and water and ensures adequate shelf life of the contents. The inner polyolefin layer, which is generally polypropylene, provides the best heat-sealing medium. This is also an inert material and hence does not materially affect the contents. The inner and outer layers are laminated using a thermosetting adhesive, conventionally a polyester-isocyanate. Schematic representation of a 3-ply pouch is shown in Fig. 9.11.

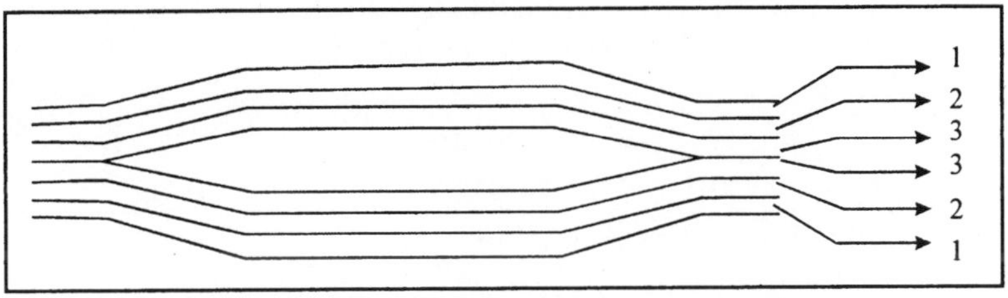

(1) polyester; (2) aluminium foil; (3) polypropylene

Fig. 9.11 : Profile of 3-ply pouch

The pouches may be fabricated on-line by the processor or purchased pre-made. The normal design is of a flat rectangular shape with four fin seals about 1 cm wide. The size is determined by the pouch thickness that can be tolerated at the correct filling weight. Generally the corners of the pouches are rounded to avoid damage to adjacent packs. A tear notch is usually incorporated for the ease of opening. The normal size ranges available are :

A 1	130 × 160 mm
A 2	130 × 200 mm
A 3	130 × 240 mm
B 1	150 × 160 mm
B 2	150 × 200 mm
B 3	150 × 240 mm
C 1	170 × 160 mm
C 2	170 × 200 mm
C 3	170 × 240 mm

The advantages and disadvantages of using the retort pouch in thermal processing of foods are the following.

9.5.4.1 *Advantages*

- because of the thin cross sectional profile heat transfer is rapid. There will be 30-40 per cent saving in processing time. No overheating of the product takes place near the package walls.
- because of the reduced exposure to heat, loss of colour, flavour or nutrients is minimum.
- shelf life is equal to or better than that of foods in metal cans.
- the empty pouches require only very little storage space compared to empty cans.
- pouches are easy to open.

9.5.4.2 *Disadvantages*

The main disadvantages are :

- pouches and pouch seals are more vulnerable to damage than are cans and hence require individual overwrap.
- with an overwrap the cost of the pouch may become higher than that of metal cans.
- slow rate of production. In place of handling 300-400 cans per minute the pouch line can handle only 30 pouches per minute.
- needs special retorts for thermal processing.

9.5.5 Glass Containers

Though glass was the original container used in heat processed foods, it has, more or less, been completely replaced by other types of containers by now. The modern glass container is a mixture of oxides viz., silica (SiO_2), lime (CaO), soda (Na_2O), alumina (Al_2O_3), magnesia (MgO) and potash in definite proportions. Colouring agents and strength improvers are added to the mixture and fused at 1350 to 1400°C and cooled sufficiently quickly so that it solidifies into a vitreous or non-crystalline condition.

Glass containers which can resist the rigours of heat processing only can be used in fish canning. The glass containers are prone to breakage due to impact or other mechanical reasons, but the one occurring due to 'thermal shock' is of greater significance in canning. Thermal shock is caused by the differences in temperature between the inside and outside walls of the container. This causes different rates of expansion in the glass wall producing an internal stress. This stress can open up minute or even microscopic cracks leading to large cracks and container failure. Chances of thermal shock will be greater with higher thickness of the glass. Therefore, glass containers used in canning should have thin walls of uniform thickness on the bottom as well as walls. As a general rule, sharp contours and flat surfaces are avoided in a glass container as more failures occur in such regions. Chemical surface coatings are often applied to make the glass more resistant to 'bruising' and to resist thermal stock.

Glass containers are hermetically sealed using closures made of tinplate or aluminium, lacquered to prevent corrosion. Closures will be lined with a resilient material, usually cork or paper over which a disk of aluminium, tin, plastic or resin is placed. The method of closing involves compressing this resilient disc against the rigid glass sealing surface and maintaining it in the compressed condition by the closure cap. The cap may be screwed on, crimped on or, pushed in or on the glass container.

Processing glass containers needs special care. A necessary precaution is to reduce thermal shock. One method is pre-heating the glass container before filling hot material and avoiding transfer of hot filled containers to cold surfaces. Glass containers are sterilised in water heated by steam. The containers are loaded in the retort containing water at about the same temperature as of the contents in the glass jars. Adequate overriding pressure is provided during heating and cooling to keep the cover in place.

9.6 SPOILAGE IN CANNED FISH

Spoilage in canned foods is quite unique. Even a defect of the container like a dent or rusting on the surface may be accounted for as spoilage even though the contents may be perfectly normal. Sometimes the can may look perfectly normal externally, though the contents might have become liquefied and unfit for consumption which can be known only on opening the can. Spoilage in canned foods takes place due mainly to microbial, chemical and physical means.

Normal cans have concave or flat ends. A can showing bulged ends is called a 'swell' or 'blower'. In a 'hard swell' the ends are permanently distended. A 'soft swell ' is one in which the bulged end can be moved by thumb pressure, but cannot be moved back to the normal position. A can in which one end is bulged but can be forced back to the normal position whereupon the opposite end bulges is called a 'springer'. A 'flipper' is a can of normal appearance, but on striking against a solid object, one end flips out. Bulged cans pass through the stages of flipper, springer and soft swell to the hard swell stage.

9.6.1 Microbial Spoilage

Microbial spoilage takes place due to several reasons, the important among them being underprocessing, inadequate cooling and leaker infection.

A canned food spoiling due to the survival and growth of bacteria is under-processed. If there is gas production a 'swell' may result; or the contents may undergo acidification and liquefaction, but no gas is produced. In the latter case the can will have a normal appearance and spoilage will be detectable only on opening the can. If the food is not grossly under-processed, the spoilage may be due to a single organism. If it contains one or more of non-heat resistant organisms, the pack can be considered as grossly under-processed or that it became infected after processing.

An important manifestation of under-processing is the 'flat-sour' spoilage. Flat-sour is a condition where the contents of the can may be acidified and liquefied without any gas production. Thermophilic spore formers of the *Bacillus* spp. surviving thermal processing are incriminated against this type of spoilage. Flat sour indicates the inadequacy of thermal processing; that is the food is under-processed.

9.6.1.1 Spoilage Producing Acid and Gas

The biological spoilage evidenced by production of acid and gas is caused by mesophilic thermophiles of the *Clostridium* spp. They decompose proteins with production of malodorous compounds like hydrogen sulphide, ammonia indole, skatole etc. The spoilage is indicated by swelled containers and decomposed foods.

Thermal processing does not envisage destruction of all the thermophiles since such a process will often result in deterioration of the quality of the food. The flat-sour thermophiles multiply rapidly in the temperature range 48-70°C and if the processed cans are not rapidly cooled to a temperature below this range, it may lead to serious spoilage by thermophiles. The risk of flat-sour spoilage will persist if the cans cooled to about 35°C are stacked immediately in large blocks.

A major threat to the safety of canned foods is spoilage arising from leakage through seams which takes place during the cooling process after thermal processing. The main source of organisms is the cooling water. In many cases the post-process infection results in 'swells' which indicate that the original leak becomes sealed by the contents or that the defective seam has some valve-like action. Leakage through seams also may result in flat-sour spoilage. Gas production and acidification also are common. The type of spoilage will depend on the nature of the organism present. In a leaker spoiled can, there may be more than one type of organism present; however, at times the cans may be contaminated by only one species like a spore-forming one. This happens when the cans are cooled in chlorinated water where the bacterial content may be reduced to resistant spore-formers.

9.6.2 Chemical Spoilage – Hydrogen Swell

Cans which swell as a result of hydrogen produced due to internal corrosion of the cans is known as 'hydrogen swell'. The bulging may range from 'flipping' to 'hard swell'. Some imperfection may occur on the tin and lacquer coatings in the cans exposing small areas of iron and tin. Tin, in contact with the product, sacrificially corrodes to protect any small exposed areas of base steel plate and iron-tin alloy. The tin coating gradually disappears and the area of exposed steel increases. Eventually, a stage is reached when the steel, initially well protected, is attacked rapidly. During this process hydrogen is evolved. Ultimately enough gas is produced to distend the can ends.

Normally 'hydrogen swells' do not occur until nearly all the available tin has been consumed. However, if the steel is of poor quality, or of the wrong chemical composition, container service life may become shortened. Can corrosion, besides producing gas, may proceed to the stage of ultimate perforation.

9.6.3 Physical Spoilage

Physically induced 'spoilage' may arise from faulty retort operation, under- exhausting or overfilling. If the steam pressure is too rapidly reduced at the end of thermal processing, high pressure will be retained inside the can causing distortion on the expansion rings of the can ends. This can present an appearance of 'swell' in the cooled can. Similar distortions are possible if the cans are not properly exhausted before seaming. The excessive internal pressure set up by expansion of the contents and the entrapped gas is responsible for this situation. Overfilling of the cans also will lead to similar conditions.

High vacuum in tall cans leads to a condition where the can body is forced inwards by atmospheric pressure particularly when thin metal is used in the making of the can. This is called 'panelling'. A similar situation can arise when the cans are cooled under pressure.

9.6.4 Miscellaneous Aspects

Externally rusted cans, particularly if after removing the rust the iron plate is seen pitted, are classified as spoiled. Damaged cans, whether it is due to rough handling or caused otherwise, especially those with any deformation on the seams also are classified as spoiled. Cans with severe dents on the body also are rejected.

9.7 INFLUENCE OF CANNING ON THE QUALITY OF FOODS

Application of heat sufficient to bring about commercial sterility in foods also causes some alterations in colour, flavour, texture and nutritive value. Heating can bring about irreversible changes in several of the components of the food such as protein, fat, carbohydrates etc. impairing the nutritive value.

9.7.1 Colour

Changes take place in the natural colour of foods on canning and is a factor immediately perceptible on opening the can. The degradation of pigments may be enhanced by the presence of metal ions. The natural carotenoid pigments in shrimp undergo some degradation during the canning process.

Besides degradation of pigments, there may be development of some coloured compounds formed by heating. Non-enzymatic browning by the action of sugars and amino acids is a common phenomenon. Sugar may undergo caramelisation in the absence of amino compounds. As there may be some air in the headspace of the cans, oxidative browning of foods is another possibility.

9.7.2 Flavour and Texture

Thermal processing degrades both flavour and texture, the extent being dependent on the sensitivity of the food to heat. Many of the flavour bearing components are sensitive to high temperature. Many fish become very soft in texture. High temperature-short time exposures to heat are less damaging to flavour and texture than an equivalent low temperature long time process.

9.7.3 Proteins

Proteins are denatured by heat, coagulated and finally precipitated. Enzymes are completely inactivated. Free sulfhydril groups increase and the heated protein tastes different. Coagulation and precipitation adversely affect the protein digestibility and hence the nutritive value is impaired even though there may be no change in the amino acid composition.

9.7.4 Fats and Oils

Fats undergo hydrolytic and oxidative rancidity. Enzymic hydrolysis produces free fatty acids. The lipolytic enzymes produced by bacteria are destroyed by heat, but oxidative rancidity is accelerated by heat, metal ions and moisture. For every 10°C increase in temperature the rate of oxidation of fat is doubled. Flavour reversion in unsaturated fat is accelerated by heat. Fat is stable to moist heat in the absence of oxygen and hence fats remain relatively stable if there is good vacuum in the can.

9.7.5 Carbohydrates

Carbohydrates undergo caramelisation at higher temperatures. Browning type reactions at high temperatures degrade carbohydrates.

9.7.6 Vitamins

Fat-soluble vitamins A and D are relatively stable to heat, but appreciable loss will occur if heated in presence of oxygen. Among water soluble vitamins thiamin and ascorbic acid are heat-labile and riboflavin stable to heat. High temperature-short time processes are less destructive to water soluble vitamins.

9.8 SOME PROBLEMS COMMONLY ASSOCIATED WITH CANNED FISHERY PRODUCTS

9.8.1 Struvite Formation

Some canned marine product such as brine packed shrimp, crab or tuna shows the presence of some glass-like crystals, particularly when the temperature of storage is low. This takes place due to the formation of a chemical compound, magnesium ammonium phosphate hexahydrate, $MgNH_4PO_4.6H_2O$, called 'struvite'. Struvite is a harmless, colourless, odourless, transparent chemical; however, it has a glass-like structure and its presence will be mistaken for fragmented glass and hence is considered objectionable in the product. Magnesium from the salt or seawater used in various operations combines with ammonia generated from the fish muscle protein during heat processing, and phosphates in the fish to form struvite and the product gradually crystalises, particularly when the cans are cooled at a very slow rate after processing.

9.8.2 Sulphide Blackening

Iron suphide blackening is generally met with in canned shrimp, lobster, crab etc. Though the fish cans are coated with a sulphur-resistant lacquer, any imperfection in the lacquer coating or any scratch on it during handling can expose tin. Trimethylamine present in marine fish will dissolve the tin from the tin layer of the container exposing iron. The sulphur-containing constituents released from the fish during thermal processing will react with iron producing iron sulphide which is black in colour. This reaction takes place readily if alkaline conditions have developed in the raw material due to spoilage. Uniform lacquering of the can and its careful handling avoiding any possible occurrence of exposea iron, maintenance of proper acidity, use of parchment lining etc. can control the occurrence of this phenomenon.

9.8.3 Curd and Adhesion

'Curd' is precipitated protein often found on top of canned fish like salmon which is generally canned without pre-cooking. Curd may adhere to the can surface and the lacquer may even get peeled off when the curd is removed. Curd is a brine soluble protein that exudes and coagulates. Use of raw fish which is not very fresh, as also inadequate brining or pre-cooking are some of the reasons found responsible for formation of curd in the can. Curd formation can be prevented by cold blanching the fish in 10-15 per cent brine for 20-30 minutes followed by washing.

9.8.4 Blue Discolouration

Blue discolouration is usually associated with canned crabmeat. Meat from the parts of the body having poor blood circulation such as legs, claws etc. shows high incidence of blueing. The copper in the haemocyanin in the crab haemolymphs reacts with the sulphur compounds liberated during thermal processing producing blue copper sulphide. This phenomenon becomes evident when the copper in the meat is above 2 mg per cent (wet weight). Therefore, the important method to prevent this consists in thoroughly bleeding the meat so that copper is reduced below this critical level. Thorough bleeding of the carcasses of crab from which the carapace has been removed in running water significantly reduces the copper content below the critical level. Another method suggested is a 'low temperature and fractional heating'. This is based on the observation that blood proteins in the crab coagulate at 69-70°C whereas meat proteins coagulate at 59-60°C. Thus, if the crab carcasses are heated to 59-60°C, the meat protein coagulates but the uncoagulated blood can run out thus bringing down the level of copper. Using a parchment lining inside the can, use of a chelating agent in the brine and maintenance of proper acidity can control discolouration in the meat.

9.8.5 Honeycombing

This is a phenomenon observed in canned tuna meat processed from stale raw material. The meat in such cans resembles honeycombs. During steaming the volume of meat contracts due to removal of water and coagulation of protein. Coagulation starts in the surface. When the water in the inner parts evaporates and escapes as gas or bubbles through the soft, not yet coagulated parts, the gelatinous parts swell like soap bubbles. The swelled gelatinous parts solidify after cooling and exhibit honeycombed appearance.

9.8.6 Softening in Shrimp

Canned shrimp often becomes very soft in cans. The cause of this softening is the decomposition of the protein to soluble non-protein components which usually occurs in the raw material when freshness declines. To prevent this strictly fresh raw material should be used for processing and high level of sanitation should be maintained in the cannery.

9.8.7 Mush

This phenomenon is a flabby condition met with in some species of pilchards caught at the end of its spawning season. It is caused by the invasion of the parasitic protozoan *Chloromyxum* which decomposes the fish meat during storage such that it becomes entirely soft during canning.

9.8.8 Retort-Burn

'Retort-burn' is usually associated with canned shellfish like clam, mussel or oyster.

This is a condition which develops when the filling medium is not sufficient to cover the solid food and the top is left 'dry'. This can be overcome by using sufficient filling medium to cover the solids in the can.

Suggested Reading

Bail, O.C. and Olson, F.C.W. (1957) *Sterilization in Food Technology*. McGraw Hill Book Co. Inc., New York and London.

Broek, Van Den, C.J.H. (1965) Fish Canning. In : *Fish as Food* Vol. IV. Borgstrom, G. (ed.), Academic Press Inc. New York.

Charm, S.E. (1978) *Fundamentals of Food Engineering*. 3rd edn. AVI Publishing Company, Westport, Connecticut, USA.

Desrosier, N.W. (1959) *The Technology of Food Preservation*. The AVI Publishing Co., Westport, Connecticut, USA.

Gillies, M.T. (1971) *Seafood Processing*. Noyes Data Corporation, London.

Gutterson, M. (1972) *Food Canning Techniques*. Noyes Data Corporation, London.

Hersom, A.C. and Hulland, E.D. (1980) *Canned Foods*. Churchill Livingstone, London.

Kyle, R.J. *et. al.* (1961) *Small Canning Facilities*. Publications & Technical Services Branch, Communications and Resources Division, Washington.

Lopez, A. (1975) *A Complete Course in Canning*. Vol. I & II *The Canning Trade*, Baltimore, Maryland.

Myers, R.R. and Long, J.S. (eds) *Treatise on Coatings* Vol. 4. *Formulations* Part - I . Marcel Dekker, New York.

Parker, M.E. and Litchfield, J.H. (1962) *Food Plant Sanitation*. Reinhold Publishing Corporation, London.

Tanikawa, E. (1971) *Marine Products in Japan*. Koseisha-Koseikaku Company, Tokyo.

Warne, D. (1988) *Manual on Fish Canning*. FAO Fisheries Technical Paper No.285 FAO, Rome.

Wheaton, F.D and Lawson, T.B. (1985) *Processing Aquatic Food Products*. John Wiley & Sons, New York.

10
FREEZING

AMONG the various methods of preservation of fish employing low temperature, icing and chilling can maintain fish fresh only for a very limited period. Quality deterioration can take place quickly in chilled fish. Most, if not all, methods of long-term preservation bring about major changes in the physical, chemical, textural and organoleptic characteristics of fish. Given a choice, consumers always prefer fresh to processed fish, or fish with minimum changes in its intrinsic characteristics. Freezing is a method of low temperature preservation of fish that can ensure very long shelf life and can also provide a processed product very much similar to fresh fish. When frozen and stored properly, fish can retain its prime quality for very long periods. In such cases it may even be difficult to distinguish frozen and thawed fish from fresh fish.

It might be the finding that foods exposed to the cold of the winter resulted in their freezing and preserved them over long periods that led to the beginning of freezing as a method of preservation of foods. Ice-salt mixture was the first man-made freezing system used for food freezing. However, freezing became prominent only with the advent of mechanical refrigeration.

Commercial freezing involves cooling to very low temperatures such that the water in the fish is converted into ice. Lowering the temperature will reduce the chemical and enzymic reactions which can lower the rate of autolysis, delay the breakdown of ATP consequently delaying the onset of rigor mortis, and also delay the bacterial spoilage. This forms the basis of preservation of foods by freezing.

The rate at which autolytic and microbial spoilage takes place in fish are dependent on the temperature of storage. Deteriorative changes involving microbial and enzymic activities can be retarded by lowering the temperature of storage and, when the temperature is sufficiently low they can almost be completely stopped. Fish preserved by freezing are normally cooled to –20°C or lower at which temperature the microorganisms and enzymes are rendered mostly inactive. Freezing binds water into ice crystals making it not available to the bacteria for their metabolic purposes.

10.1 FREEZING FISH

The major constituent of fish, and of any living cell, is water. Water accounts for 75- 80 per cent of the weight in most fish. At normal atmospheric pressure water will freeze at 0°C. Water in the fish contains several dissolved organic and inorganic substances like sugars, salts and other compounds. Besides these, more complex organic molecules like proteins are also present as colloidal suspension. Freezing involves reducing the temperature of the body to 0°C or below leading to crystallisation of a part or whole of the water and some of the solubles.

Freezing point of a liquid is the temperature at which its vapour pressure is in equilibrium with that of the solid. The vapour pressure of a solution is the sum of the partial pressures of the solute and the solvent. A solution with vapour pressure lower than that of the pure solvent will not be in equilibrium with the solid solvent at its normal freezing point. So the suspension will have to be cooled to a temperature at which the solution and the solid solvent have the same vapour pressure. Dissolved solids bring about a lowering in the freezing point of the solvent. This is true of foods also. The freezing point of any food will be lower than that of pure water. Freezing begins in the fish usually at -1 to -2°C.

10.1.1 Freezing Process

10.1.1.1 Principles of Refrigeration

In the simplest terms refrigeration means removal of heat from a body or atmosphere, which is desired to be cooled, and its transfer to another medium. If the cooling medium is available at a low temperature, e.g. ice, the operation can take place automatically by virtue of the tendency of heat to flow from a higher to a lower temperature level. When it is necessary to transfer heat from a comparatively cold body to a warmer one special methods have to be adopted. Vapour compression refrigeration system is the most commonly employed system in freezing and frozen storage of fish and fishery products. This system makes use of (i) the large capacity of a liquid to absorb heat, i.e., the latent heat necessary to convert the liquid into vapour, and (ii) the relationship between pressure and temperature of a saturated vapour.

Refrigeration is accomplished by the evaporation of a liquid refrigerant, which extracts heat from the medium to be cooled. The refrigeration cycle involves steps to remove this heat from the evaporating refrigerant by again converting it into the liquid state in order that it may be used repeatedly in a continuous process. For this it is necessary that the gas be compressed permitting condensation at the normal temperatures of the available water or air. This is done by compressors requiring mechanical energy.

A schematic representation of the refrigeration cycle is given in Fig. 10.1. It consists of a compressor, a condenser, a receiver, a throttle valve known as expansion or regulator valve, and an evaporator. The refrigerant maintained as a liquid under high pressure is admitted to the evaporator through the expansion valve. The expansion valve maintains the necessary pressure difference corresponding to the required difference of temperature between the condenser and evaporator sides of the circuit. The liquid refrigerant while passing through the expansion valve evaporates due to the fall in pressure. The latent heat necessary for evaporation is taken from the liquid which becomes cooled to a temperature corresponding to the reduced pressure. The refrigerant entering the evaporator, therefore, is in the form of cold liquid containing a certain proportion of vapour. The liquid takes up its latent heat from the surroundings of the vapour and leaves the evaporator in vapour form. The vapour coming out of the evaporator is compressed in the compressor under pressure when its temperature is raised sufficient to enable it to give up the heat to the cooling medium of the condenser. It leaves the condenser as a liquid and enters the receiver. The cycle is repeated.

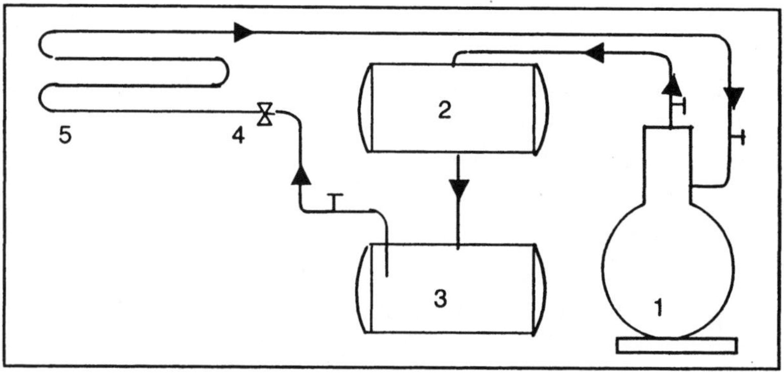

(1) compressor; (2) condenser; (3) receiver; (4) expansion vale; (5) evaporator

Fig. 10.1 : Vapour compression system

10.1.1.2 Quantity of Heat to be Removed for Freezing Fish

Freezing involves removal of heat from the product that takes place in distinct stages. In the first stage specific heat is removed from the product lowering its temperature to a stage when freezing begins. Specific heat is the heat to be removed from 1 g of water to lower its temperature by 1°C. The total heat energy to be removed from the body during this stage is :

$$H_1 = S_1 \times W (T_i - T_f),$$

where H_1= kJ of energy to be removed from the body to lower the temperature from initial T_1 to the temperature at which it freezes T_f,

S_1 = Specific heat of water above freezing point;

W = Weight of fish in kg.

Once freezing begins a change of state of water from liquid to solid takes place. The temperature will remain more or less steady during this period. Latent heat of fusion will have to be removed from the fish to change water into ice. Latent heat is the amount of heat which must be removed at a constant or relatively constant temperature to change the state of material, e.g., liquid to solid. The heat energy to be removed at this stage is

$$H_2 = H_f \times W$$

where H_2 = kJ of energy to be removed to change the fish at freezing point to solid state at freezing point.

H_f = Latent heat of fusion of fish

W = Weight of fish in kg

In the third stage the temperature of frozen fish is lowered to the temperature at which it is stored. The heat to be removed is

$$H_3 = S_2 \times W (T_f - T_s)$$

where H_3 = kJ of energy to be removed to lower the temperature from freezing point (T_f) to temperature of storage (T_s)

S_2 = Specific heat of frozen fish

W = Weight of fish in kg

Therefore the refrigeration requirement is equal to $H_1 + H_2 + H_3$ kJ. In real arithmetic terms, considering that a kg fish at 30°C is to be frozen to –30°C, the following calculation will apply :

Specific heat of water	4.2 J (1.0 cal)
Specific heat of ice	2.1J (0.5 cal)
Latent heat of fusion of ice	334.7 J (80 cal)

For the purpose of these calculations it is assumed that fish has the same specific and latent heat as of water

Stage 1 Cooling from 30°C to -1°C

$$1000 \times 31 \times 4.2 = 130200 \text{ J} = 130.2 \text{ kJ}$$

Stage 2

$$1000 \times 334.7 = 334700 \text{ J} = 334.7 \text{ kJ}$$

Stage 3

$$1000 \times 29 \times 2.1 = 60900 \text{ J} = 60.9 \text{ kJ}$$

Total energy to be removed = 130.2 + 334.7 + 60.9 = 525.8 kJ

However, in actual practice it will be still greater because of the necessity to remove heat from the packaging material, glaze etc.

10.1.1.3 Freezing Curve

Temperature profile of freezing fish is presented in Fig. 10.2. Initially the temperature falls rapidly until it reaches just below 0°C (stage 1) while specific heat of water is removed. Though not apparent in all cases, some super cooling occurs in many cases, where the temperature falls below the normal freezing point without any formation of ice. When crystallisation commences, a rise in temperature takes place due to release of heat of crystallisation and the temperature reaches the initial freezing point. During this stage the bulk of the water changes to ice and therefore the temperature remains more or less constant at –1 to –5°C (stage 2). This period is known as the 'thermal arrest time'. During the early phases of thermal arrest time water separates as pure or nearly pure ice crystals. During later stages eutectic mixtures and other complex solids may form. At the end of the thermal arrest time the specimen will contain much less freezable water. Therefore removal of relatively smaller amount of heat energy will cause a much greater reduction in the temperature (stage 3). At this stage most of the remaining water freezes though the specimen may still contain some freezable water. However, at very rapid rates of heat removal the various stages described above may become less distinct and even indistinguishable.

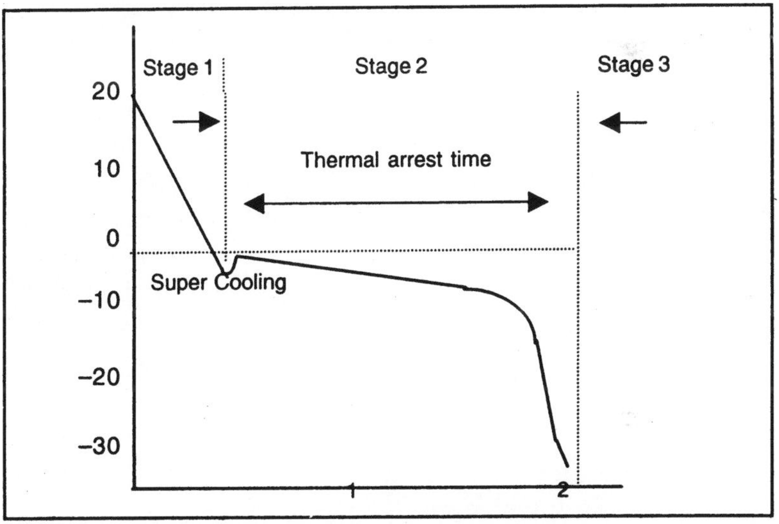

Fig. 10.2 : Typical fish freezing curve

10.1.1.4 Water Frozen at Different Temperature Ranges

Generally, crystallisation of water in fish commences when the temperature is lowered to –1°C. Percentage of water frozen at different temperature ranges are given in Table 10.1.

Table 10.1 : Percentage of Water Frozen at Different Temperature Ranges

Temperature range □C	per cent of water frozen
–1	Crystallisation begins
–3	70
–5	80
–25	About 90
–50 to –60	Almost all

It can be seen that the largest portion of water freezes between –1 and –5°C. The rate of cooling at this temperature interval determines the size of ice crystals formed. This temperature zone which also corresponds to the thermal arrest time is also called the "zone of maximum crystallisation".

10.2 FACTORS AFFECTING FREEZING TIME OF FISH

In the freezing process, basically heat is transferred from one medium to another. Heat is removed from the product mainly by conduction and convection. Heat from the centre of the product to its outside surface is transferred by conduction and is a function of thermal conductivity, temperature difference and thermal properties of the product. Heat transfer to the cooling medium takes place by convection or radiation.

Freezing rate will depend on several factors such as method of heat transfer and temperature difference between the fish and the cooling medium. Size, shape and thermal properties of the fish, packaging materials used and the method of packaging are some of the other factors involved.

Freezing is directly related to the temperature difference between the product and the cooling medium. Freezing time decreases in direct proportion to increase in difference in their temperature. Freezing time can be considered directly proportional to the square of the product thickness.

Packaging materials, depending on their thermal

conductivity, can increase or retard transfer of heat. It is desirable to use thin packaging materials having high thermal conductivity that can allow rapid heat transfer. Packaging should be tight and air pockets should be avoided.

10.3 QUICK FREEZING *VS* SLOW FREEZING

There is no internationally accepted definition for quick freezing or slow freezing though these terms are commonly used while referring to food freezing. It is already known that the largest proportion of water in fish freezes between –1 and –5°C, i.e., the zone of maximum crystallisation. The rate of cooling at this zone also determines the size of ice crystals and some other biochemical reactions affecting the quality. Therefore, it is desirable that the duration of this zone should be as short as possible. In the light of this, the freezing process in which the fish passes through the zone of maximum crystallisation in 30 minutes or less is arbitrarily referred to as quick freezing. Another general expression of rate of freezing is in terms of the average speed at which ice front moves in the material during freezing and is expressed as millimeters per hour. Based on this the nature of freezing, whether quick, slow etc., can be expressed as in Table 10.2.

Table 10.2 : Type of Freezing and Rate of Movement of Ice Front

Type of freezing	Rate of movement of ice front, mm/h
Slow freezing	2 or less
Quick freezing	2-30
Rapid freezing	30-100
Ultra rapid freezing	More than 100

Freezing rate may also be better applied to the time taken for the centre of the product to attain a certain temperature. Quick freezing is the process employed in commercial freezing of fish. Quick freezing is preferred to slow freezing because

- during slow freezing large ice crystals are formed in the cells of the fish. The ice crystals can be larger than the cells themselves and, hence, can break the cell walls. This will lead to texture changes and increased thaw drip.

- water tends to freeze out as pure water leading to a higher concentration of salts and enzymes in the unfrozen water. This can accelerate autolysis.
- some bacteria will remain live at around 0°C causing bacterial spoilage to continue in the fish.

10.4 FREEZING METHODS AND EQUIPMENT

10.4.1 Direct and Indirect Systems

Refrigeration systems can be broadly classified into two, the direct expansion system and the indirect system. In the direct expansion system, the refrigerant absorbs heat directly from the material to be cooled. Therefore the evaporator coil is placed directly in the place to be refrigerated. In the indirect or brine system, the refrigerant absorbs the heat that the brine absorbs from the material to be cooled. The place to be refrigerated is therefore cooled by cold brine which has been previously cooled by the refrigerant system which is in operation at some place other than the place of operation. Both these systems are commercially employed for freezing different products.

Freezing methods commonly employed for seafood freezing can be broadly classified into

- freezing in air
- indirect contact freezing
- immersion freezing
- cryogenic freezing

10.4.2 Freezing in Air

Air is the most common freezing medium. Air freezing systems are of two types - still air freezing and forced air freezing

10.4.2.1 Freezing in Still Air

The freezer consists of an insulated room or a cabinet maintained at –28 to –45°C. Fish, packaged or otherwise, placed in aluminum trays, is kept on shelves made of pipes or coils through which the refrigerant is circulated. The time taken for freezing may be 12 hours or more. Freezing time is influenced by such factors as the temperature of the freezer, size and shape of fish, presence or absence of packaging, load in the freezer and the arrangement of fish in the freezer. Freezing in still air is the least expensive method of freezing; however, is the slowest method as well.

10.4.2.2 Air Blast Freezer

Air blast freezer consists of an insulated room or a tunnel. Air is cooled by blowing by a fan through the finned cooling coil of the refrigeration system. Cool air passes over the fish to be frozen and picks up heat from the product, walls of the freezer etc. Temperature is maintained at –35 to –40°C. Fish, packaged or otherwise and kept in trays or suitable containers, is placed in the air blast. Fish gets frozen by being in contact with the vigorously moving cold air.

The main advantages of air blast freezing are :

- a versatile method of freezing applicable to any size and shape of fish
- advantageous for producing individually quick frozen fish, fish fillets or coated products
- suitable for operation as batch or continuous process

The main disadvantages are :

- they occupy more space compared to other types of freezers
- they consume more energy
- unpackaged fish will undergo excessive dehydration
- fish will swell giving a bulged appearance if not confined between rigid surfaces during freezing

10.4.2.3 Continuous Air Blast Freezer

An improvement in air blast freezing is to use a conveyer to move the fish continuously through the room or the tunnel. The speed of the conveyor can be varied to suit the type of fish being frozen. Air flow can be either countercurrent to the movement of the material or across the belt. Air velocity maintained at 150–300 m/sec makes intimate contact with the fish and accomplishes freezing at a fast rate. Freezing is quicker and any type and size of fish can be frozen in large quantities in shorter periods. Spiral belt freezers in which a long belt can be compacted into a relatively limited space is an example of this type of freezers.

10.4.2.4 Fluidised Bed Freezing

Fluidised bed freezer is an improvement over the continuous belt freezer. Fluidisation is a method of keeping solid particles partially supported in a rising column of cold air. The particles kept on a mesh belt are individually suspended in an upward stream of cold air at a velocity sufficient to float the particles in the air. Each particle is surrounded by air and is kept suspended separated from each other. In this stage the particles will assume the properties of a liquid. Freezing is rapid as the best possible heat transfer is ensured between the cold air and the product. Air velocity of 120 m/min and operating temperature of –35 to –40°C are common in fluidised bed freezing. This technique is most suitable for small sized and uniform products like shrimp, fish fillet, small fish etc. Advantages of fluidised bed freezing are

- rapid rate of freezing due to very efficient heat transfer
- less product dehydration and consequently less frequent need for defrosting the equipment (dehydration loss is only 1 per cent in fluidised bed freezing of shrimp)
- short freezing time generally responsible for low moisture loss

- no distortion of shape; the original appearance is not lost

The major disadvantage is that, large and non-uniform products cannot be fluidised at reasonable air velocity.

10.4.3 Indirect Contact Freezing

Indirect contact freezing may be defined as freezing a product by keeping it in contact with a metal surface which is cooled by some refrigerant. In this system the refrigerant does not come into direct contact with the material being frozen. It is only circulated through hollow plates which are refrigerated and absorb heat from the product placed on it either directly or in metal trays. These freezers are called contact plate freezers and are of two types - the horizontal plate freezer and the vertical plate freezer. In both types the product is brought into close contact with metal plates through which the refrigerant is circulated. The freezers are equipped with hydraulic systems to move the plates closer or apart. By this arrangement the product can be compacted to make closer contact with the plates for quicker freezing and also to release the product quickly after freezing. Multiplate double contact plate freezer is the version most commonly employed in fish freezing.

10.4.3.1 Horizontal Plate Freezer

Horizontal plate freezers are used in almost all land based freezing operations. These freezers, in general, have 15–20 plates. The product to be frozen contained in metal freezing trays are loaded between freezing plates and are maintained in close contact with top and bottom plates under slight hydraulic pressure to ensure maximum heat exchange. The freezing trays are also covered with tightly fitting lids which will help to provide adequate contact with freezing plate at the top. Temperature is maintained between –35 and –40°C. Fish freezes in 2–2½ hours.

The freezing trays may be lined with polythene prior to packing the fish. Fish can be held in cardboard cartons as well. Depending on the product the freezing trays may be divided into compartments to produce frozen blocks of uniform size. Trays and cartons should be filled to the top with the material to ensure good contact of the product with freezing plates at the top and the bottom.

Horizontal plate freezing provides the following advantages :

- produces well shaped uniform blocks
- floor area requirement is the minimum
- freezing is quick
- under good operating conditions the product dehydration is the minimum and hence the freezer does not require defrosting

The main disadvantages, however, are :

- needs much handling in loading and unloading
- freezing rate will become slow if air spaces are there in the package

10.4.3.2 Vertical Plate Freezer

Vertical plate freezers are most commonly used for freezing fish at sea. They consist of a number of vertical freezing plates that form partitions in a container. The spaces between the plates are known as stations. Temperature ranges between –30 and –40°C. Fish are dropped between the plates until each station is full. The plates are then closed together to form fish blocks. Fish of similar size should be packed in each block.

Contact plate freezing is a very economical method. Dehydration taking place in the product is minimum and consequently there is practically no need for defrosting. The product remains in uniform blocks without any bulging.

10.4.3.3 Rotary Drum Freezer

Rotary drum freezer is a refrigerated stainless steel drum revolving at a pre-set speed. Freezing time can be adjusted by adjusting the speed of revolution. Unpacked products like fish fillets, shrimp etc. can be rapidly frozen using rotary drum freezer. The material to be frozen is fed on the external surface of the drum through a conveyor. It adheres immediately to the surface of the drum by freezing the water on the surface of the material. At the end of one revolution the frozen product is scraped off and is passed through an automatic glazer before packaging. As there is no movement of air and freezing is very rapid, there will be little or no loss of weight during freezing.

10.4.4 Immersion Freezing

In this method, freezing is achieved by immersion in, or spraying with, a refrigerant that remains liquid throughout the process. Refrigerated aqueous solutions of propylene glycol, glycerol, sodium chloride, calcium chloride and mixtures of sugars and salt can be used as the medium for freezing. Immersion freezing allows for intimate contact of every surface of the material with the freezing medium and thus ensures very efficient heat transfer.

10.4.4.1 Freezing in Brine

The most common in the immersion freezing category employs refrigerated aqueous solution of sodium chloride. The solution can be made to any strength to achieve the desired freezing point. Saturated brine freezes at –21°C; however, in practice the operating temperature cannot be reduced below about –18°C. Further reduction in the temperature of the fish, if desired, can be achieved by transferring the product to cold storage maintained at the required temperature.

Brine freezing is rapid and is adaptable for continuous operation. However, the fish will absorb some salt the extent of which will depend on several factors including the temperature of the brine, duration of immersion, fat content in the fish and the surface area. Salt absorption can, to a great extent, be reduced by using mixture of glucose or corn syrup and salt as the freezant. The glucose-salt solution will also provide a protective glaze to the product and hence the frozen product will not stick together. Contamination of the cooling medium leading to cross contamination of the subsequent batches of material to be frozen is a major disadvantage of the process.

10.4.4.2 Brine Spray Freezing

Instead of immersion in refrigerated brine, freezing can also be achieved by spraying the fish with refrigerated brine. Chilled brine is sprayed over the fish placed in trays. Heat is removed from the fish by the chilled brine eventually freezing the fish in 1-2 hours.

10.4.5 Cryogenic Freezing

In cryogenic freezing very rapid freezing is achieved by exposing the fish, unpackaged or with a very thin package, to an extremely cool freezant undergoing a change of state. The essential difference between cryogenic freezing and liquid immersion freezing is the change of state in the former while removing heat from the body. The most common food grade cryogenic freezants are boiling nitrogen and boiling or subliming carbondioxide.

Cryogenic freezing is much faster than air blast or contact plate freezing; but is only moderately faster than fluidised bed or liquid immersion freezing. For example, shrimp is frozen in nine minutes in a commercial liquid nitrogen freezer whereas it takes 12 minutes to freeze in a fluidised bed and 1-2 hours in contact plate and air blast freezers.

10.4.5.1 Freezing Using Liquid Nitrogen

In freezing using liquid nitrogen, liquefied nitrogen gas is sprayed over the product as it travels in a tunnel over a conveyor belt. The nitrogen gas passes countercurrent to the movement of the fish so that the fish gets pre-cooled before reaching the liquid nitrogen spray. After the spray the product is allowed to temper for a while before being discharged from the tunnel. A schematic presentation of a cryogenic freezing system employing liquid nitrogen is presented in Fig. 10.3.

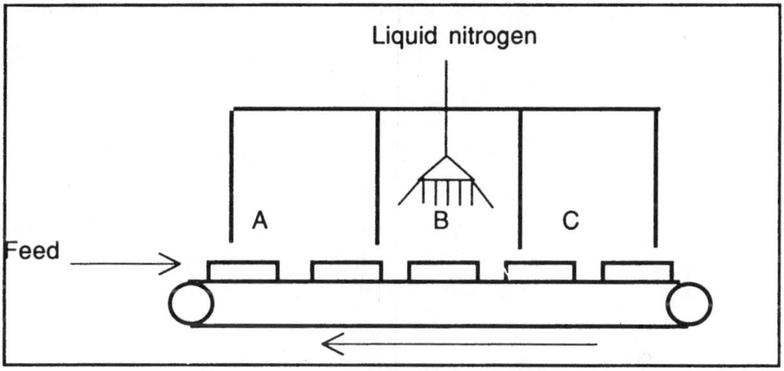

Fig. 10.3 : Liquid nitrogen freezing

The freezer is an insulated chamber divided into three zones. The product placed on a conveyor belt is moved on into the zone 'A' where it is pre-cooled by the modestly cool nitrogen vapour running countercurrent. 'B' is the freezing zone where liquid nitrogen is sprayed onto the product. After the required exposure it is moved to zone 'C' where it is allowed to equilibrate to the desired final temperature.

Advantages of freezing with liquid nitrogen over other methods are :

- negligible dehydration loss.
- oxygen is excluded in the process and hence the product is protected against any oxygen induced changes during storage.
- attains individual quick freezing of the product and freezing damage is minimum.
- equipment is simple. It is adaptable to various production rates and product sizes, is suitable for

continuous operation and high product turnout can be achieved with minimum space and time.
- initial investment is low.

10.4.5.1 Freezing Using Liquid/Solid Carbon Dioxide

In this type freezing, liquid carbon dioxide is sprayed over the fish as it travels through a tunnel on a moving conveyor. As the fish passes under the nozzles its surface is sprayed with carbon dioxide in the liquid form. While spraying through the nozzles, the pressure of liquid carbon dioxide is quickly reduced and about 50 per cent of it suddenly changes into tiny snowy particles which absorb large quantities of heat from the surroundings and gets converted to gas resulting in rapid cooling of the fish. Freezing can also be achieved by exposing the products to powdered solid carbondioxide.

Freezing with liquid/solid carbon dioxide has most of the advantages of freezing using liquid nitrogen. However, unpackaged foods may absorb or entrap carbon dioxide to the extent of causing undesirable swelling.

Liquefied nitrogen and solid carbon dioxide are also used as supplementary cooling agents to maintain low temperature in the fish holds while transporting frozen cargo. Solid carbon dioxide is used in transportation of frozen fish by air.

10.4.6 Freezing Using Liquid Refrigerant

The most common liquid refrigerant used is dichlorodifluromethane (Freon-12). Fish placed in a mesh belt is conveyed to an insulated chamber. Fish is frozen either by spraying the product with food grade Freon-12 or by a combination of initial immersion in liquid Freon followed by spraying with it. In both cases the vapours are collected for reuse.

The process has all the advantages of liquid nitrogen freezing and the added advantage of cost. Even though

Freon-12 boils at a much higher temperature than liquid nitrogen, the freezing rates are more or less similar with both. However, owing to concerns over the effects of Freon-12 in ozone depletion in the atmosphere its use has rapidly declined.

10.4.7 Double Freezing

It is a common practice to preserve fish by freezing onboard vessels when the voyage lasts for few weeks. On reaching the shore the fish is thawed and reprocessed. Use of fish frozen in bulk onboard for further processing into fingers, reprocessing block frozen shrimp into individually quick frozen shrimp etc. are the common examples of double freezing.

Quality-wise, re-freezing of fish is considered undesirable. Lean fish, as fillet, suffers considerable loss in quality, particularly in texture that becomes tough. Fatty fish, however, withstand such treatment better.

10.4.8 Partial Freezing

Partial freezing, or superchilling, involves lowering the temperature of the fish to about -2 to -3°C. By cooling the fish to this temperature range, approximately half of the unbound water present in the fish will get converted to the solid phase. Storage life of the partially frozen fish is almost double that of the fish stored in ice. However, some loss of sensory quality is experienced when the storage period extends to five days. This is considered to be caused by the fluctuations in the storage temperature. A change in temperature of less than 0.5°C can be instrumental in intermittent melting or freezing of water in the fish and can have pronounced influence on protein denaturation. Post-mortem glycolysis resulting in lowering the pH of the muscle and resultant decrease in the water holding capacity of the fish protein also may contribute to protein denaturation. Storage life of partially frozen fish, if subsequently frozen, is very short.

Maintaining strictly constant temperature in the storage is very essential to avert loss of quality in partially frozen fish.

10.4.9 Cold Storage of Frozen Fish

The frozen fish has necessarily to be cold-stored for some period before shipping or consumption. Correct design and proper operation of cold stores are absolutely necessary to maintain the product quality. Dehydration or 'freezer burn' is the most common problem encountered in fish during cold storage. Though the fish is frozen to very low temperatures of even upto –40°C, the cold storage is often maintained only at around –20 to –25°C. The problem of freezer burn arises principally from a change in the relative humidity in the cold store. It is known that warm air can hold more water than cold air. When air warms without the addition of water vapour, the relative humidity drops. Warming of air can take place when an unfrozen or partially frozen product is introduced into the cold storage, or when there is ingress of air due to the doors being frequently opened. Warming can also take place due to leakage of air through door frames or other fittings. The decrease in the relative humidity increases the risk of freezer burn in the product.

An important defect in the cold storage atmosphere often possible is non-uniform distribution of temperature. Circulation of air will then cause the product to have varying temperatures at different locations leading to increased dehydration. Similar problems will persist if the cut-off and on temperatures are set too far apart.

Air circulation, while is necessary to maintain uniform temperature in the cold storage, should be maintained at the minimum necessary, as excessive circulation can cause increase in the incidence of dehydration. Air, as it passes over the product, can pick up moisture and hence dehydration will be more with more air passing over it. Similarly the design should ensure that the heat ingress, though it is unavoidable, is minimised.

10.5 CHANGES ASSOCIATED WITH FREEZING AND COLD STORAGE OF FISH

Freezing is considered a very gentle method of preservation. The influence of freezing on many foods is so little that it will be practically impossible to distinguish between an unfrozen and, frozen and thawed fish. Freezing can be generalised as consisting of crystallisation of water and some of the solutes. Significance of important factors contributing to this phenomenon and their influence on quality of fish are discussed below.

10.5.1 Supercooling

Supercooling is a condition attained by a liquid where its temperature is lowered to below its freezing point without crystallisation taking place. When the temperature of a biological system is reduced to sub-zero level, the solution first 'supercools' before solutes start crystallisation. Supercooling, in general, need not impair the quality of the foods. However, this need not necessarily be the case in freezing fish; quality impairment may take place depending on the conditions of freezing. The undesirable side effects of freezing are considered to be due to the water-ice transformation rather than reduction in temperature.

10.5.2 Crystallisation

Crystallisation is the process of formation of a well-defined solid phase from a liquid or solution. Two different stages in crystallisation are nucleation and crystal growth.

Nucleation is of two types – homogeneous nucleation taking place in pure liquids and heterogeneous or catalytic nucleation taking place in foods and living specimens. The latter involves formation of nuclei adjacent to suspended solid particles, surface films or walls of containers.

In aqueous systems the temperature invariably falls temporarily below the freezing point (supercooling) before nucleation and further solidification can occur. As temperature is lowered to a critical level characteristic of the sample, nucleation begins. Once this occurs any further decrease in temperature will result in an abrupt increase in the rate of crystallisation.

Crystal growth can take place at temperatures just below the melting point of the system. Growth of ice crystals occurs as a result of the water molecules adding to the existing nuclei. Growth rate of crystals increases moderately with decrease in the rate of heat removal.

10.5.2.1 Location of Ice Crystals

Location of ice crystals is governed by the freezing rate, temperature of the specimen and nature of cells. During freezing of food products, formation of ice crystals takes place first in the liquid between the cells. This is presumably due to the higher freezing point of the extracellular water compared with that of the intracellular water. During slow freezing ice crystals are formed exclusively in the extracellular locations. Instances of intracellular ice formation also have been detected. Slow freezing leads to preferential extracellular freezing resulting in large ice crystals and maximum dislocation of water.

Formation of ice crystals causes concentration of the solutes in the extracellular liquid. The cells will lose water by osmosis leading to extensive dehydration of cells and causing the cells to contract. The result will be few, but relatively large ice crystals occurring between the cells. The food quality of the product also will remain reduced.

During rapid freezing ice formation takes place more or less uniformly in extracellular and intracellular water. Heat is removed so quickly that there is little time for dehydration of the cells and ice will be formed in the cells also. Rapid freezing causes uniform crystallisation and produces numerous small ice crystals. There is minimum dislocation of water and the frozen appearance of the product is quite similar to that of the original unfrozen specimen. The food quality also is graded superior to that obtainable in a slow frozen product.

From the above discussions it can be concluded that a slow frozen product will lose some amount of water during thawing because the water will not be able to return to its original position. Such problem is not likely to arise in the case of a rapidly frozen product.

10.5.3 Volume Changes

Density of ice is 0.92 g/ml. Therefore, pure water when transformed into ice at 0°C expands approximately by 9 per cent. On freezing, most foods expand but to a lesser extent than that of pure water. Volume increases of food as a result of ice formation is about 6 per cent. There are exceptions; for example, highly concentrated solution of sucrose exhibits a small contraction in volume during freezing. Volume change during freezing is influenced by the following factors.

10.5.3.1 Composition of the Food

Water is principally responsible for all expansions. However, change in volume may be insignificant when the material contains intracellular air space that can accommodate growing crystals and minimise the changes in the exterior dimensions.

10.5.3.2 Fraction of Water that Fails to Freeze

Bound or supercooled water does not contribute to expansion. The nature and amount of solutes present influence the amount of bound water and tendency of the system to overcool.

10.5.3.3 Temperature Range

Many of the changes taking place in the systems due to lowering the temperature have different influences on the volume of the product. Water expands to approximately

9 per cent whereas most of the constituents contract as the temperature is lowered and therefore the volume changes will not be uniform. There will be localised areas of expansion (like formation of ice crystals) and localised areas of contraction. This will result in development of stress with the possibility of mechanical damage to the product.

Contraction during cooling prior to freezing, expansion during ice formation, contraction associated with solute crystallisation, contraction during cooling of non-solutes such as fat etc. are some of the specific changes in volume associated with progressive cooling.

10.5.4 Concentration of Non-aqueous Constituents

During freezing water is converted into ice crystals of high degree of purity. As a result of this, concentration of non-aqueous constituents in the unfrozen water will increase. This is a process more or less similar to the conventional dehydration except that this operation is carried out at sub-zero temperatures and that the water separated is deposited locally as ice instead of being removed from the system as in dehydration. The increased concentration of solutes brings about significant changes in pH, titratable acidity, ionic strength, viscosity, freezing point, surface and interfacial tensions and, oxidation-reduction potential. Other probable changes are expulsion of oxygen and carbondioxide, alteration in the water structure and water-solute interaction, forcing together of macromolecules providing opportunity for detrimental interactions and formation of eutectics.

During freezing, concentration of solutes may exceed their saturation point and simultaneous crystallisation of ice and solutes become possible. The temperature at which a crystallised solute can exist in equilibrium with ice and unfrozen phase is called the 'eutectic point' or 'eutectic temperature' of the solute. Maximum ice formation is not possible until the eutectic temperature of that constituent with the lowest eutectic temperature is reached. This temperature is referred to as the final eutectic temperature. Natural foods have eutectic temperatures in the range –55 to –70°C. This will explain why all frozen foods processed and stored at conventional temperatures contain some unfrozen and freezable water. Some bound and unfreezable water also may be present.

Slow rate of removal of heat, as in slow freezing, results in smooth continuous solid liquid interface. This also results in maximum crystal purity and maximum concentration of solute in the unfrozen phase. The ice crystals formed are large sharp needle-like structures. The large sharp crystals can penetrate the cell walls and rupture them causing large drip loss during thawing. However, it is now fairly well understood that the important quality changes associated with freezing and thawing are caused by the irreversible changes in the colloidal state of the proteins induced by the increased concentration of solutes in the unfrozen phase. Walls of fish muscle are elastic enough not to be excessively spoiled by large ice crystals. Quality loss is more related to the denaturation of proteins, which is influenced by. temperature, concentration of enzymes and other compounds present in the muscle. When water freezes out, the concentration of enzymes continues to increase in unfrozen water, thus increasing the rate of denaturaion. Denaturation is maximum at –1 to –2°C. Therefore it is essential that fish should pass through this temperature range during freezing as also thawing as quickly as possible.

Rapid removal of heat results in an irregular or discontinuous solid-liquid interface, considerable entrapment of solutes by growing crystals and less than maximum concentration of solutes in the unfrozen phase. Ice crystals formed are minute in size and will have a fine texture.

10.5.5 Thaw Drip

Thaw drip is the exudate of tissue fluids that flow free from fish muscle during thawing. Denaturation of the myofibrillar proteins results in a tough, dry and fibrous fish muscle. In the latter stages raw, thawed fish becomes dull and opaque in appearance, and the structure becomes spongy. In this condition the raw thawed fish tends to lose fluid. The volume of drip will depend on the type of the product, freezing method employed, conditions of storage etc. Drip may vary from 3 to 5 per cent and even upto 20 per cent of the total fish weight following prolonged storage. The somewhat dry texture experienced in some thawed fish can be considered to be due to drip loss. Drip formation takes place as a result of cell damage as also due to dehydration of protein micelles taking place as a result of the denaturation of proteins.

10.5.6 Gaping of Fillets

On thawing, frozen fish fillets often exhibit a phenomenon in which the connective tissues fail to hold the muscle blocks together exhibiting 'gaps' in the musculature. This phenomenon observed in fish which are well-fed at the time of capture and are frozen before or during rigor is called 'gaping'. The major cause of gaping is the weakening of the connective tissues which bind together the muscle segments in the fish flesh. Weakening of the connective tissues results in the muscle segments falling apart as in cooked fish flesh. Gaping can partly be explained as due to the muscle contraction during freezing which may tend to pull the segments apart and thus show a tendency to divide along the myocommata.

10.5.7 Thaw Rigor

Muscle of the fish frozen prior to the onset of rigor mortis will contain a sizeable proportion of undegraded ATP. The muscle of such fish, when thawed, can contract so strongly that a high proportion of tissue water is pressed out from the fillets as drip. In exceptional cases this may reach upto 30-50 per cent. This phenomenon is called 'thaw rigor'.

10.5.8 Recrystallisation and Desiccation

Major physical changes associated with frozen storage are recrystallisation and desiccation. The size and shape of the ice crystals produced during freezing are relatively unstable. They undergo metamorphic changes, i.e., changes in the number, size, shape, orientation or perfection, after the initial solidification and is collectively termed 'recrystallisation'. The quality advantages of quick freezing will disappear partly during storage if recrystallisation takes place in the system. The rate of recrystallisation will decrease at lower, constant storage temperatures.

Desiccation or sublimation of ice takes place from the surface of unpackaged or improperly packaged foods causing what is termed 'freezer burn'. Fish develops a chalky white, yellow or brown colour and wrinkled appearance on the surface. When thawed, such fish presents a dry and spongy texture. Greater temperature differences between the fish and the evaporator and greater velocity of circulating air cause high degree of moisture loss from the surface of the fish. Moisture loss will be more at higher storage temperatures. Fluctuations in the storage temperature will influence the weight loss as well as the quality of fish. This can be minimised by any means which can effectively stop loss of moisture such as maintenance of high relative humidity in the storage, use of glazes, use of packaging materials with low or very low water vapour permeability and providing a tight fitting package with no air space in between.

10.5.8.1 Glazing

In order to prevent desiccation the most important precaution taken is to glaze the fish. In products like block frozen shrimp, water is added to the pack before freezing. After freezing, the frozen blocks are again dipped in ice-cold water for a few seconds or chilled water is sprayed over them. This provides a uniform thin coating of ice all around the block. Glazing, in addition to preventing dehydration, also helps in preventing the oxidative rancidity in fish. The amount of glaze pick up will depend on

- temperature of the fish.
- temperature of glaze water.
- size and shape of the product.
- glazing time.

10.5.9 Discolouration

Some frozen fish like tuna exhibit green or brown discolouration on cooking. Browning is due to the formation of metmyoglobin in the muscle through autoxidation of ferrous myoglobin. Greening takes place due to the presence of pigments resulting from the oxidation of haemochrome that occurs when the meat is unduly exposed to oxidative conditions during and after cooking. Proper evisceration and bleeding of fish immediately after catch reduces the risk of discolouration. Loss of characteristic pink colour in shrimp, lobster etc. is the direct consequence of the changes in the naturally present carotenoid pigments like β-carotene and astaxanthin.

10.5.10 Toughness

Freezing and cold storage results in toughness of fish meat and this, as also dryness on cooking, increases during cold storage. The increase in toughness is partly due to the changes in proteins like denaturation and cross-linking reactions of the myofibrillar proteins, and also due to desiccation. It is also influenced by the state at which the fish is frozen. Freezing fish in rigor is generally undesirable. However, it is less important when whole fish is frozen. In the case of fillets it is desirable that rigor mortis be completed at a non-freezing temperature prior to freezing in order to minimise contraction of muscle and lesser increase in toughness.

10.5.11 Protein Denaturation

The most important change associated with freezing and frozen fish is denaturation of protein. The structural changes taking place in the protein molecules bring about definite changes in the chemical, physical and biological properties of the protein. Curdling of proteinaceous materials observed especially during repeated freezing and thawing is an important physical observation associated with protein denaturation.

Denaturation results in decreased solubility of protein. Reduction in the amount of salt soluble protein is, therefore, often taken as a criterion of the extent of denaturation. It has been reported that during frozen storage the extractability of the myofibrillar group of proteins decreases whereas the extractability of the sarcoplasmic proteins remains unaffected. The loss in the characteristic texture due to decrease in water holding capacity of myofibrillar proteins brings about increased drip loss in some species of fish.

Reasons for changes in these protein fractions may be the localised concentrations of salts throughout the flesh resulting in salting out of proteins. Fish flesh undergoing this type of deterioration often has the appearance of salted undried product of the same kind.

The lipid constituents of the fish muscle have an important influence on the stability of proteins. Changes in the protein can be correlated with accumulation of free fatty acids or formaldehyde produced as a result of the cleavage of trimethylamine oxide to dimethylamine and formaldehyde. The rate of protein denaturation also depends on several factors like storage temperature and the fluctuations taking place therein, state of rigor of the fish at the time of freezing, rate of freezing and thawing etc.

Another change that can take place in the fish muscle protein during storage is proteolysis if the enzymes present have not been inactivated.

Deteriorative protein changes in fish can be prevented or minimised by employing very low and constant temperature in the storage and by inhibiting the oxidation of lipids and desiccation of the product.

10.5.12 Glycolysis

On death, glycogen in fish is converted into lactic acid resulting in a fall in pH. The glycogen content in fish may generally be low because the methods employed for fishing often lead to its exhaustion in the fish before death. The fall in pH is often to the tune of 6.4 to 6.8. Fish flesh with low pH has a tendency to gape, and to develop toughness during frozen storage.

10.5.13 Changes in Fat

Most fish contain fat in their body, the proportion being dependent on several factors like species, size, maturity etc. Fish lipids contain a much higher proportion of poly-unsaturated fatty acids than meat lipids. Lipids in frozen fish tend to become rancid quicker than those in other animal tissues.

Two types of rancidity can develop in fish lipids, hydrolytic rancidity due to the degradation of triglycerides to glycerol and free fatty acids; and oxidative rancidity leading to the formation of hydroperoxides which will break down to acids and carbonyl compounds responsible for rancid flavours and odours. Oxidation of fat may cause changes in colour of foods, especially the carotenoid pigments. The peroxides formed as a result of fat oxidation can oxidise the unsaturated carotenoids causing a shift in the absorption peak towards the ultraviolet end of the spectrum.

Fish with high oil content are the most susceptible to oxidative changes, but this depends also on the degree and proportion of unsaturated fatty acids in the oil and presence or absence of natural antioxidants like tocopherols and pro-oxidants like salt or heme compounds.

Development of rancidity is also related to the temperature of storage. It is very fast at -15°C, significant at -20°C and very slow at -35°C. Both types of rancidity, hydrolytic and oxidative, will be accelerated by a rise in the storage temperature or by the presence of metallic ions, particularly of copper and iron. However, the enzymic rancidity is very limited at the commercial frozen storage temperature of -20°C or less because the enzymic activity largely ceases at temperature below -15°C. Besides maintaining low temperature of storage, the other methods to control rancidity include proper glazing, use of packaging materials having very low oxygen permeability and use of antioxidants like butylated hydroxy anisole (BHA), butylated hydroxy toluene (BHT), propyl gallate etc.

10.5.14 Influence on Microorganisms

Microorganisms may be classified depending upon the optimum temperature for their growth. Growth of microorganisms at temperatures below the freezing point of fish depends upon the water activity at the temperature concerned. When a great amount of water is frozen the growth of microorganisms become exceedingly slow and lag period will extend to weeks or even months. It is implied that such temperatures will considerably reduce the growth of microorganisms. If the temperature is exceedingly low microorganisms may not grow at all. However, yeast and, moulds can grow at temperatures much lower than those supporting the growth of bacteria.

Many bacteria are killed during freezing and the survivors die off slowly. Because of the long lag periods of bacterial growth at temperatures considerably below 0°C the product may not undergo any serious deterioration in quality due to microbial growth. However, frozen fish is not free from viable bacteria and other microorganisms and is, therefore, subject to fast spoilage when defrosted and held at temperatures suitable for their growth.

Destructive effect of freezing process on the micro-organisms also depends on the pH of fish, type and extent of contamination, temperature of freezing and storage etc. The growth will be much less at lower pH than at neutral. Heavier the initial load, greater will be the number of survivors. Microorganisms of the Gram-positive type can survive freezing while others cannot, and freezing does not seem to change their properties. Vegetative cells are less resistant to freezing than are spores. Fat, sugar and colloidal materials in the frozen media often protect the bacterial cells against destruction by freezing.

Food poisoning bacteria are known not to grow under conditions of freezing. Enterotoxic staphylococci do not grow below 10°C and salmonella below 5°C. *Clostridium botulinum* types A and B do not grow and produce toxin below 10°C, however, type E has been found to grow and produce toxin at temperature as low as 3°C.

There is some controversy regarding the effects of rate of freezing on the microorganisms. However, it is generally explained that slow freezing appears to be more effective in destroying bacteria than rapid freezing, but slow freezing is not recommended because during the pre-freezing periods the bacteria get time to grow before the temperature of the food becomes sufficiently low to inhibit their growth resulting in undesirable effects on quality, particularly texture and flavour.

In fish, in general, freezing causes an initial reduction in the bacteria present to the tune of 60-90 per cent. If the temperature of storage is maintained below the minimum for growth there will be further drop in their number during storage.

10.5.15 Influence on enzymes

Enzyme activity is dependent on the temperature of

storage. It is related to an optimum pH and is influenced by the concentration of the substrate. Enzymes are inactived at high temperatures, but the activity is only retarded by freezing temperatures. Enzyme activity is stimulated when the food remains in supercooled stage. Velocity of enzyme reactions is greater in supercooled water than in crystalline water at the same temperature.

Partial crystallisation of water increases the concentration of enzymes and other solutes in the unfrozen phase. There is much uncrystallised water present in the system at temperatures above –10°C. Hence, holding the fish above –10°C, but below its freezing point leads to severe damage in the quality of fish. Damage occurs in the appearance as well as nutritive value. Long term storage at such temperatures makes the fish unacceptable. Enzymic activity will almost cease if the temperature is lowered to –15°C or lower.

10.5.16 Influence on Vitamins

Freezing as such does not bring about any deterioration of vitamins and this is of minor interest with respect to freezing of fish. Of particular interest in fish are the fat-soluble vitamins and they are known to be altered little by freezing.

10.5.17 Storage Temperatures

Thawed frozen fish resembles fresh fish. In order that a fish more or less similar in characteristics to those of the unfrozen fish is obtained, the chemical changes leading to alterations in the quality should be retarded or suppressed to the maximum possible extent. It is now well understood that most of the irreversible changes take place in the range –1 to –5 °C and hence temperature warmer than –10°C are avoided in frozen storage.

Storage of frozen fish at –10°C had been considered satisfactory. This will arrest microbiological activity. However, the concentration of solutes in the unfrozen phase and consequent pH changes and enzyme activities can definitely accelerate some irreversible changes causing severe damage to the quality of fish.

Frozen fish stored at –20°C has almost four times the shelf life compared to that stored at –10°C. Frozen fish stored at –30°C has almost double the shelf life that obtainable at –20°C. For long time storage of frozen whole fish, fish blocks or fish products the ideal temperature is –20° to –30°C or lower.

10.6 THAWING

Thawing is the reverse of the freezing process. When heat energy is added to the frozen fish the ice is converted into water and fish takes up the original state. Thawing involves addition of latent heat of fusion through a stationary level of water, which enlarges with time and raising the temperature of the product. In the reverse process of freezing the latent heat of crystallisation is removed through an ice layer which enlarges with time and the product temperature is lowered. The thermal conductivity of ice is about four times that of water. As a result, the thawing time will substantially exceed the freezing time under identical conditions, and if heat is supplied to the surface of the product.

During thawing the product undergoes an initial rapid rise in temperature caused by the temperature differentials between the external ice layer in the product and thawing atmosphere. This causes the surface ice layer to thaw forming a layer of non-flowing water on the surface. Further heat transfer from the product becomes significantly impaired because of the poor thermal conductivity water. The relatively longer time taken for thawing and the temperature differentials within the product make the food susceptible to chemical, physical and microbial spoilage and hence thawing has to be considered as a greater potential source of damage than freezing. The thawing system should be so selected that it avoids localised overheating of the fish, excessive drip loss, dehydration and bacterial growth.

The differences in the rates of freezing and thawing are important when, in the thawing process, heat energy is transmitted by conduction. This will not apply in the process involving dielectric or microwave heating for thawing.

Thawing systems can be divided into two general groups; systems depending on conduction of heat to the centre of the product and that depending on some method other than conduction. Thawing in air, water and vacuum depends on conduction. Dielectric thawing, electrical resistance thawing and microwave thawing are the important methods under the second group.

10.6.1 Conductive Thawing Methods

10.6.1.1 Thawing in Air

Thawing can be done in still air or in an air blast. Thawing in air blast can be batch or continuous process. In the continuous process cold air, usually at 15-20°C, is blown across or at right angles to the conveyor carrying frozen fish. Air thawing favours product dehydration. Therefore, the air used must be as nearly saturated as possible with moisture. Small fish, poultry etc. are generally thawed in air.

10.6.1.2 Thawing in Water

Water has better heat transfer properties than air. Therefore, thawing by immersion in water reduces the thawing time compared with that in air. Frozen fish is immersed in water in tanks or water is sprayed over the fish. Temperature of water should not be high enough to be detrimental to the quality of fish. It is also important that fish should be prevented from absorbing too much water. Large fish, fish blocks and similar materials are thawed in water. Thawing in water may, sometimes, become detrimental to the quality of fish in respect of flavour, taste and appearance.

10.6.1.3 Vacuum Thawing

After placing the material in a chamber, partial vacuum is drawn and then steam is injected into the chamber. Condensation of steam liberates 2.45 kJ of heat energy per kg of steam to the product thawing it rapidly.

Alternately, instead of injecting steam, water placed in a tray on the base of the chamber may be heated to fill the chamber with water vapour. The heat liberated by the condensing vapours will be absorbed by the fish.

10.6.2 Non-conductive Methods of Thawing

10.6.2.1 Electrical Resistance Thawing

The product to be thawed is placed between two stainless steel plates and an electric current is passed through them. Electrical resistance of fish decreases with increasing temperature. Thus warmer the block, the more the current passing through it and heating the product faster. However, this method is suitable only for uniform products. With uneven packs, the portion getting good electric contact will heat up more rapidly and may result in cooking in this portion. Hence this method is not suitable for whole fish or whole fish blocks.

10.6.2.2 Dielectric Thawing

The frozen material is placed between two metal plates such that the plates do not touch the material. The plates are connected to a high voltage high frequency alternating current. Motion of charged particles in the product generates heat energy in the material and causes rapid thawing.

If the frozen product is reasonably homogeneous dielectric thawing is more rapid and much more rapid than in conduction heating. However, fish is not remarkably homogeneous because of the non-uniform distribution of lipids and will tend to localised overheating leaving some areas frozen while others are thawed.

10.6.2.3 Microwave Thawing

This employs frequencies in the 3000 MHz range to heat the product. Due to its high content of water, fish is an excellent microwave absorber and heating occurs throughout the product. It is ten times more rapid than dielectric thawing.

10.6.2.4 Hybrid Thawing Methods

Hybrid thawing methods involve the use of combination of various thawing process, e.g. the first one-third in water, and the rest two-thirds by direct heating.

10.7 PACKAGING REQUIREMENTS FOR FROZEN FISH

Packaging is a major factor in storage and transport of frozen fish. Prevention of desiccation in frozen fish is greatly achieved by proper packaging. Even if it is agreed that proper and adequate glazing is the simplest and most effective protection against desiccation, the need for further packaging to ensure undue accumulation of frost on freezer plates or cold stores should be emphasised. The packaging materials used for frozen fish should have the following desirable qualities.

10.7.1 Mechanical Strength

In order that a food contained by it is physically protected, the package must maintain its mechanical integrity. Frozen fish is stored at low temperatures and there is always the possibility of some moisture getting condensed over the package. To offset any loss in strength brought about by such contingencies, the packaging used should have high mechanical strength. They should have mechanical properties like good tear and burst strength at low temperatures as also high 'wet strength' i.e., high tensile strength under conditions in which the material is saturated with water. Packaging materials used should be such that they should protect the contents against normal transit and storage hazards.

10.7.2 Flexibility

The package is subject to stress and strain due to the expansion of frozen food and changes in the moisture content of the package itself. The product is usually packed without headspace to avoid the presence of air. Therefore the package used should be flexible enough to conform closely to the shape of the product.

10.7.3 Liquid Tightness

The package should be capable of preventing any passage of liquids This requirement assumes importance since many frozen foods are thawed in the packages themselves.

10.7.4 Water Vapour Permeability

The most important problem in freezing and storage is desiccation. Hence, one of the major functions of the packages is prevention of desiccation. It is of great significance that the paçkages should not permit passage of water vapour. This will be possible if the packaging material has very low water vapour permeability. Even using a packaging material of low water vapour permeability care should be taken to provide a tight package without allowing any air inside.

10.7.5 Permeability to Gases and Vapours

Frozen fish, because of its fat content is susceptible to oxidative deterioration during storage. Therefore another important quality requirement of the packaging materials used is that they do not permit passage of gases like oxygen which can affect flavour, keeping quality etc. of the product. The permeability to flavours and odours also is of equal importance. Many of compounds responsible for flavours and odours have higher vapour pressure, even at low temperature, and can migrate rapidly through the packaging materials. Loss of flavour as well as pickup of undesirable odours from the outside will be the consequence of such migration.

10.7.6 Impermeability to Light

Deterioration taking place in foods is often catalysed by light. Discolouration of pigments as in the case of astaxanthin in crustaceans like shrimp or lobster, myoglobin of red meat, oxidative rancidity in fat etc. are accelerated by light. Therefore the package used should preferably be impermeable to light.

10.7.7 Migration of the Components of Packaging Materials

Yet another consideration in the selection of packaging materials is that there should be no migration of package components into the food. Such components, in many cases, impair the safety of foods. This becomes all the more important when the packaging employed is the 'boil-in-bag' type.

Suggested Reading

Bald, W.B.(ed.) (1991) *Food Freezing: Today and Tomorrow.* Springer-Verlag, London.

Brown, D.C. (1966) Liquid nitrogen freezing in the shrimp industry. *Quick Frozen Foods.* 29 : 111-112, 142-143.

Charm, S.E. (1971) *The Fundamentals of Food Engineering.* AVI Publishing Company, New York.

Desrosier, N.W. (1959) *The Technology of Food Preservation.* The AVI Publishing Co. Inc., Westport, Connecticut, USA.

Desrosier, N.W. (ed.) (1977) *Fundamentals of Food Freezing.* Tata McGraw Hill Publishing Co. Ltd., New Delhi.

FAO (1977) *Freezing in Fisheries.* Fisheries Technical Paper No.167, FAO, Rome.

Heid, J.L. and Joslyn, M.A. (1981) *Fundamentals of Fish Processing Operations.* The AVI Publishing Co. Inc., Westport, Connecticut, USA.

Jul, M. (1984) *The Quality of Frozen Foods.* Academic Press, London.

Kreuzr, R. (ed.) (1969) Freezing and Irradiation of Fish. *Fishing News*, London.

Robinson, R.K.(ed.) (1985) *Microbiology of Frozen Foods.* Elsevier Applied Science Publishers.

Tressler, D.K. and Evers, C.F. (1957) *Freezing Preservation of Foods.* Vol. I & II. The AVI Publishing Co. Inc. Westport, Connecticut, U.S.A.

Tressler, D.K. van Arsdel, W.V. and Copey, M.J. (1968) *Freezing Preservation of Foods.* Vol. III. The AVI Publishing Co. Inc. Westport, Connecticut, U.S.A.

11
MODIFIED ATMOSPHERE PACKAGING

CONSUMERS always prefer fresh fish to preserved/ processed fish, if the former is readily available. Though chilling and freezing are two methods of preservation/ processing which are known to bring about least changes in the intrinsic qualities of fish and where the product is nearest to fresh fish in characteristics, they lack the 'convenience' factor. Most of the other methods of processing bring about extensive changes in several respects in the fish. There is also an increasing demand for fresh and chilled 'convenience foods' containing few or no preservatives. A method of extending the shelf life of foods that can more or less meet the above demands of the consumers is packaging them in modified atmosphere. Modified atmosphere packaging (MAP) can result in significant increase in the shelf life without loosing many of the attributes of the 'fresh' fish. The MAP can be described as the process by which the shelf life of a food product is increased by enclosing it in an atmosphere so modified that it slows down the degradation processes such as the growth of microorganisms and development of oxidative rancidity. Retention of the colour and other physical attributes of certain foods are other benefits associated with MAP. Many food products such as meat, poultry, fish, vegetables etc. are now packed in modified atmosphere. Technological developments in the production of several polymeric films having high and wide ranges of gas permeability have provided an added impetus to the development of MAP as a commercial success.

The MAP can extend shelf life of foods without loss of many of the 'freshness' attributes. The MAP helps to retain at least some of the properties of the living tissues for their existence in 'fresh' state, e.g., fruits and vegetables. With highly perishable foods like fish, though the above may not hold good, shelf life extension even by only a few days with the 'freshness' attributes retained will be of great commercial significance.

11.1 METHODS OF MODIFYING THE ATMOSPHERE FOR PACKAGING

Any change brought about in the atmosphere in which a product is maintained results in some modification in the atmosphere. The types of modifications that are generally brought about in the packaging of foods, particularly fish and fish products, are :

- modified atmosphere.
- controlled atmosphere.
- vacuum packaging (hypobaric storage).
- hyperbaric storage.

11.1.1 Modified Atmosphere Packaging (MAP)

This is employed mostly in the case of non-respiring products, e.g., meat, cheese, fish, baked goods etc. In modified atmosphere packaging, the air in the pack of food, fish or fish products for example, is replaced with a mixture of gases. Carbondioxide, nitrogen and oxygen are the gases generally used in modifying the atmosphere. The composition of the gases will vary depending on the nature of the product. The proportion of each component gas will be fixed for a given product at the time of introduction of the gas mixture into the pack. However, there will be no further control of any sort on the composition of the gases. It cannot be expected that the composition of the atmosphere at the time of sealing will be maintained throughout the storage period even if the packaging film is intact and properly sealed. The gas mixture may undergo thorough changes during the period of storage because of the gases diffusing into and out of the package at different rates as well as the gases being absorbed or given off by the product. Biochemical activity also may modify the atmosphere.

11.1.2 Controlled Atmosphere Packaging (CAP)

In controlled atmosphere packaging also, a mixture of gases of known composition is introduced into the pack. The essential difference of CAP from MAP is that in the former, the composition of the gas inside the package is continuously controlled throughout the period of storage. This is achieved by including within the package chemical agents of various sorts. To prevent the product from contamination these agents are packed in sachets permeable to gases being scavenged or produced by these agents. The CAP technique is mainly employed in bulk storage of products.

11.1.3 Vacuum Packaging (Hypobaric Storage)

Exclusion of the air from the package and thus creating a vacuum also is, in effect, a certain type of modification of the atmosphere. Vacuum packaging represents a static form of hypobaric storage which is widely used in the food industry due to its effectiveness in reducing oxidative reactions in the product at relatively lower costs. In vacuum packaging, the product is contained in a package made of a material having low oxygen permeability and is sealed airtight after evacuating the air. Vacuum packaging has been shown to extend the shelf life for periods varying from six days upwards. However, even though the product may not develop rancidity in extended periods of storage, it may develop objectionable odours and flavours due to bacterial activity. The little gaseous atmosphere likely to be present in the beginning in the package will undergo changes during storage because of the metabolism of the product and/or action of microorganisms and all the volatiles produced as a result of the decomposition are sealed within the package.

11.1.4 Hyperbaric Storage

Hyperbaric storage refers to storage using high-pressure systems. High pressure can stop microbial growth and reduce enzyme activity. There is experimental evidence that refrigerated storage of lean fish at high pressure extends the shelf life considerably as determined by bacterial counts and sensory evaluation. It is widely accepted that conformational changes of protein takes place at high pressure which may be responsible for shelf life extension. In many surimi-based products gelling is an important function and fish muscle protein paste forms a gel on application of high hydrostatic pressure. Thus, application of high pressure helps to formulate a number of products with good functional properties. However, because of the technical difficulties in building a commercially feasible high-pressure storage unit, this method of preservation has not become popular.

11.2 ROLE OF GASES IN SHELF LIFE EXTENSION

Individual gases play their role in extending the shelf life of the MAP products. However, it is almost always a mixture of gases employed and their composition will be decided on the basis of the product in the package.

11.2.1 Carbondioxide

Carbondioxide is the major gas employed in the modified atmosphere packaging of fish and fishery products. Carbondioxide has powerful inhibitory effect on the growth of bacteria and mould when present at or above 20 per cent. Some anaerobic and facultative bacteria also are inhibited by carbondioxide and hence this effect cannot be attributed to the presence of oxygen alone in the pack. The overall effect of carbondioxide on microorganisms manifests in an increased lag phase, and a decreased growth rate during the logarithmic phase of bacterial growth. Inhibitive action of carbondioxide on microorganisms is more effective when the product is stored at lower range of refrigerated temperatures.

Carbondioxide is soluble in lipids and water. High concentration of carbondioxide may lead to the production of carbonic acid in the fish and this will lower the pH of the muscle, produce an acid taste, reduce the water holding capacity of the proteins and hence, may result in the formation of drip in the pack. High concentrations of carbondioxide also produce organoleptic changes like coarsening of the texture and alteration of the colour of the belly flaps, cornea and skin of the fish. Excessive absorption of carbondioxide by high moisture/high fat foods like fish or meat will lead to lowering of internal pressure in the pack and becomes a potential reason for the phenomenon known as 'pack collapse'. Because of the above reasons use of carbondioxide in high concentration in the fish package is generally avoided.

11.2.2 Nitrogen

Nitrogen is an inert, tasteless gas with low solubility in lipids and water. It is mainly used to displace oxygen in the packs so that the onset of oxidative rancidity can be delayed and inhibit the growth of aerobic microorganisms. It can also be used as a filler gas to prevent any likely pack collapse in packs with high concentrations of carbondioxide. Nitrogen also helps to prevent mould growth and insect attacks.

11.2.3 Oxygen

Oxygen will inhibit the growth of strictly anaerobic bacteria, but will stimulate the growth of aerobic bacteria. Oxygen is usually excluded from the atmospheres used for fatty fish. Oxygen may be included for lean fish packs for considerations of anaerobic packaging. However, in the case of some foods like respiring fruits inclusion of oxygen in the atmosphere is necessary for the basic metabolism. If the level of oxygen is maintained below 2 per cent in the atmosphere, the development of rancid odours and flavours can be greatly retarded.

It can be seen that each of the above gases individually can have beneficial as well as negative effects on the products. The success of the process will depend on developing the correct composition of the gas mixture where the negative effects are balanced against their positive contributions. The optimum levels have to be worked out for individual species of fish and fish products in relation to their intrinsic characteristics.

11.3 MODIFIED ATMOSPHERE PACKAGING OF FISH

11.3.1 Composition of Gas Mixture

Various combinations of gases have been studied for their effect on extension of shelf life of fish. One of the widely accepted combination for lean fish is 40 per cent carbondioxide, 30 per cent nitrogen and 30 per cent oxygen. For fatty fish, a combination of 60 per cent carbondioxide and 40 per cent nitrogen has been found most suitable. The temperature of storage also significantly affects the shelf life. With the above combination of gases extension of shelf life achieved at 0°C is up to nine days for lean fish, whereas at 2°C it is only five days.

11.3.2 Effect of MAP on Pathogenic Bacteria in Fresh Fish

The MAP is known to extend the shelf life of fish mainly because of the sensitiveness of aerobic microorganisms to carbondioxide. Gram-negative bacteria are more sensitive to carbondioxide than Gram-positive bacteria, and therefore, psychrophiles that are more prominent in storage at chill temperatures generally get inhibited. However, concern exists in the safety hazards of MAP fish because of the possibility of allowing or enhancing the growth of pathogenic organisms like *Clostridium botulinum* while suppressing the bacteria causing spoilage. Other psychrotrophic pathogens of concern are *Aeromonas, Listeria* and *Yersinia.*

The most important among the bacterial hazards is from the potential growth and toxin production by the psychrotrophic strains of *C. botulinum*. They are non-proteolytic and hence can grow and produce toxin without producing any visible signs of spoilage. Spoilage may be absent even during the growth of *C. botulinim* and production of toxin because of the suppression of the spoilage bacteria by the packaging atmosphere. Strains of *C. botulinim* type E and the non-proteolytic types B and F are able to grow at temperatures as low as 3.3°C without showing obvious signs of spoilage. As the MAP does not involve any other mechanism to control these pathogens it is always advisable that hygienically handled raw material from known sources only are used for packaging under modified atmosphere.

11.3.4 Packaging

The type of packaging material used is crucial to the success of MAP. The gas mixture employed at the start of packaging the product may not remain so for long if the packaging material has poor barrier properties allowing passage of gases. More often than not, suitable barrier properties are incorporated into the packaging material using multilayer material composed of two or more co-extruded or laminated films.

The factors to be considered for deciding the appropriate package for a particular product are

- whether it should be rigid, semi-rigid, flexible or of any other type.
- barrier properties. In order that the initial composition of the gas mixture is retained to the maximum possible extent, it is desirable that packaging material with very good barrier properties is used.

- mechanical strength. The requirement of mechanical strength of the packaging material will depend on whether the product has sharp or smooth contours and also the hazards of handling and transportation it may be subjected to.
- seal integrity. While maintaining the tightness of the seal during storage and transport, it should not be difficult to open as well. The right balance between tightness and security and the easiness of opening are the factors to be considered.

Most of the developments in MAP in general are concerning the retail packs. These include the development of microwavable and resealable MAP packs. Another development is the triple-web packs which have the advantages of MAP and vacuum skin packing. However, many of the developments are yet to be extended to the MAP of fish and fish products.

11.4 FUTURE OF MODIFIED ATMOSPHERE PACKAGING FOR FISH

The use of MAP for storage and transport of fish has been projected to have a very high potential. However, it also has some major limitations and attendant cost factors. Some of these are

- need of very high quality raw material. With a highly perishable product like fish for which only a very limited extension is achieved, any loss of quality in the raw material will greatly affect the expected shelf life.
- need for very strict temperature control. It has been demonstrated with very fresh fish and ideal gas combination that the shelf life extension could be nine days at 0°C whereas it falls to five days when the temperature is 2°C. Any fluctuation in storage temperature significantly affects the shelf life.

- need of special equipment. Special equipment are needed for administration of the gas mixture into the packages, package sealing etc. Specialised training is needed for the personnel employed in the operation.
- need to determine the ideal gas mixture for individual products which will require great amount of research input.

All the above aspects, individually or collectively, add to the production costs, thus affecting the economy of the operations.

Suggested Reading

Anon. (1982) Control atmosphere packaging. *Packaging News.* (September) : 65-66.

Banner, R. (1978) Vacuum packaging for fresh fish. *Food Eng.* 50 : 92-93.

Brown, W.D., Watts, A.M., Heyer, D., Spruce, B. and Price. R.J. (1980) Modified Atmosphere Storage of Rockfish and Silver salmon. *J. Food. Sci.*, 45 : 93-96.

Cann, C.D. *Packing Fish in a Modified Atmosphere.* Torry Advisory Note No. 88. Ministry of Agriculture, Fisheries and Food, London.

Haard, N.F., Martins, I., Newbury, R. and Botta, R. (1979) Hypobaric storage of atlantic herring and cod. *Can Inst. Food Sci., Technol. J.* 12 : 84-87.

King, A.A. and Nagel, C.W. (1965) Growth inhibition of a *Pseudomonas* by carbon dioxide. *J. Food Sci.*, 35 : 375.

Mead, G.C. (1983) Effect of packaging and gaseous environment on the microbiology and shelf life of processed poultry products. In : *Food Microbiology: Advances and Prospects,* Roberts, T.A. and Skinner, F.A. (eds), Academic Press Inc., London and New York.

Nelson, R.W. and Tretsven, W.I. (1975) *Storage of Pacific Salmon in Controlled Atmosphere*. Proc. XIV International Congress on Refrigeration, Moscow, USSR.

Parkin, K.L., Wells, M.J. and Brown, W.D. (1981) Modified Atmosphere Storage of Rockfish fillets. *J.Food. Sci.* 47 : 181-183.

Wheaton, F.W. and Lawson, T.B. (1985) *Processing Aquatic Food Products*. John, Wiley & Sons, New York.

12
MINCE AND MINCE-BASED PRODUCTS

FISH mince or minced fish is the flesh separated from the fish in a comminuted form free from scales, skin and bones. Trimmings from the fish filleting operations, whether hand or machine filleting, was the original source of fish mince. In principle, flesh can be separated from any size and species of fish in the minced form. But production of mince as a new technology assumes importance when applied to low value fish, particularly the bycatch from shrimp trawling, which face difficulty in marketing as fresh fish or processing into conventional products. Significant value addition will accrue to such fish if converted into fish mince because mince can be used to process a variety of value added products having high commercial potential. Another advantage of the process is that it conceals identity of the original fish from which it is made and consumers may not hesitate to accept mince or mince based products even though the original fish would have been unacceptable as whole fish.

12.1 PRODUCTION OF FISH MINCE

Though mince can be made out of any fish, for reasons of its shelf stability and also on consideration of the quality of the products processed incorporating it, fish species low in fat and high in meat content are considered more suitable. When performed on a small scale and when relatively large fish are used, the flesh can be separated from the fish by hand filleting and mince can be made by passing the fillets through a conventional meat mincer. However, most of the fish used for production of mince are very small and available in very large quantities, e.g, the shrimp bycatch. In such cases mechanical devices of separating the flesh in the minced form are necessary. Meat/bone separators are used for this purpose. These machines separate the soft parts of the fish like flesh, guts etc. from the harder parts like bones, scales and fins. Principle of production of mince using a meat/bone separator is illustrated in Fig. 12.1

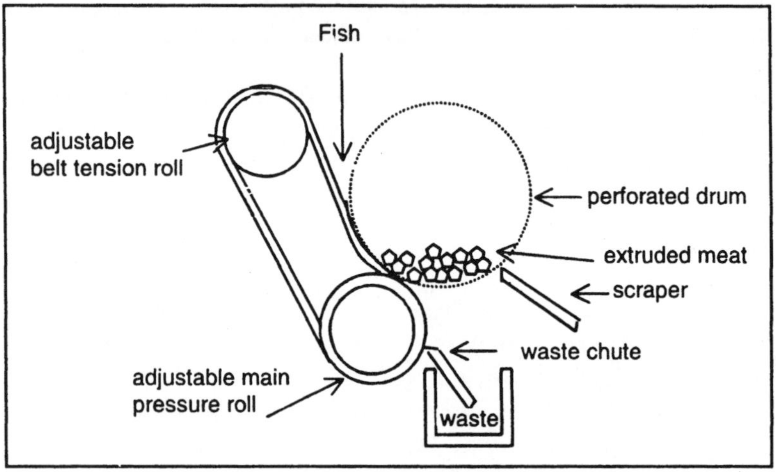

Fig. 12.1 : Design of meat bone separator

Several versions of meat/bone separators are available, but the most common types have an adjustable belt and perforated drum rotating in opposite directions, between which the fish is pressed. The flesh, with some blood and fat, is forced into the drum and, the skin and bones are rejected. The tension of the belt and the size of the perforation on the drum decide the yield and quality of mince. Size of the perforation ranges from 1-5 mm. With lower size perforation the mince obtained is of fine size and good colour; but the losses will be high during washing. The yield is better when the size of the perforation is large, but the product will need refining.

The yield of mince using a meat/bone separator will be higher than that obtained by other methods like hand/machine filleting and mincing the flesh. Meat/bone separator can recover the meat from head, belly as also from the bone frame. This will not be available when the fish is filleted and minced.

Mince can be prepared either from gutted fish or whole fish. In the latter case the mince will have a repelling colour and flavour because of the presence of gut contents, blood etc. Therefore, the fish is often gutted and washed before preparation of mince. Mince from gutted fish will be better in colour, appearance and flavour. For best quality minces it is desirable to use only a single species of fish so that mince of different quality and stability do not mix together. It goes without saying that for quality mince, quality fish should be used.

The freshness of the raw material profoundly influences the quality of the mince. The mince will contain all the components like the enymes, lipids, haem pigments etc. in the fish that affect its shelf stability. The mince is frozen and stored. While in store denaturation of myofibrillar proteins is possible, which will be more pronounced in mince than in whole fish or fillets. The intimate mixing of the components that takes place as a result of the mechanical separation of meat is responsible for the increased incidence of denaturation in mince.

12.2 MINCE-BASED PRODUCTS

The fish mince finds application in processing several 'convenience foods' like fish finger, cutlet, burger and also in some low cost salted and dried products. In preparation of fish finger, stick, cake etc., the mince stripped from the bone frame is incorporated to increase the yield.

12.2.1 Fishfinger

Fishfinger is a very popular product made out of fish mince. The mince is mixed with 1 per cent salt, made into rectangular slabs and frozen. The frozen mince is cut into suitable uniform sizes. These pieces are given a coating of batter followed by breading. The battered and breaded fishfingers are flash-fried in oil maintained at 180-200° for about 20 seconds. After cooling the fingers are frozen and stored.

12.2.2 Fish Burger

Burger is made using mince from lean white fleshed fish. Cooked mince is mixed with cooked potato and mild spices and formed into flat round pieces. These are battered, breaded and flash-fried as for fish fingers.

12.2.3 Salted Fish Cake

The mince is mixed with salt. Salting will denature the protein and reduce its water holding capacity and hence will result in release of water. The fish/salt mix is pressed suitably to release more water. The resultant cake is dried.

Several other products like cutlet, fish ball, paste etc. can be processed out of fish mince.

12.3 SURIMI

Surimi is a Japanese term for mechanically deboned fish mince from white fleshed fish that has been washed, refined and mixed with cryoprotectants for better frozen shelf life. Washing helps to remove blood, pigments, fat, soluble proteins and odour bearing substances. Consequently, the concentration of myofibrillar proteins, which are responsible for improved gel strength and elasticity, gets increased. Because of its high gel strength, surimi is used as intermediate in processing several value added products with simulated texture, flavour and appearance such as shrimp, lobster tail, scallop meat and crab leg.

12.3.1. Nature of Muscle Proteins in Relation to Surimi

Fish muscle is of two types, dark and white. Active swimmers like tuna and mackerel have a large proportion of dark muscle compared to the muscle of sluggish fish. Though both muscles are essentially similar in composition, dark muscle has a higher content of haem pigments and lipids. This will affect the colour and flavour of surimi. Surimi should have white or near white colour and bland flavour and therefore, only white fleshed fish with low fat content are preferred for processing.

Surimi is actually a wet concentrate of the myofibrillar proteins in the fish muscle which give the fish its fibre-like structure and muscular activity. Myofibrillar proteins constitute 65-80 per cent of the total proteins. The other major fraction, the sacroplasmic proteins, is water-soluble and will be lost while washing the mince during processing surimi. Neither do they contribute to gel formation. Stroma proteins or the connective issues constitute only a minor fraction of 3-5 per cent of the total fish muscle proteins. The stroma proteins are not removed by the surimi process. These are easily solubilised by cooking. The myofibrillar proteins are of extreme significance as they are responsible for the textural qualities such as fibrousness, water-holding capacity, gel forming ability and plasticity which are often referred to as the functional properties. Therefore, recovery, concentration and protection of myofibrillar proteins are the important considerations in processing surimi.

12.3.2.1 Selection of Raw Materials

Surimi is most economically made from low-value, abundant species with year round availability. White-fleshed non-oily fish like the Alaska pollack (*Theragra chalcogramma*) is the best suited for surimi production. One of the best tropical species for surimi production is threadfin bream (*Nemipterus japonicus*). Even though there are other fishes offering a gelling ability higher than that of Alaska pollack, this species was the most popular raw material for surimi because of its price advantage and abundance. However, the reduction of the stock of Alaska pollack forced the surimi processors to look for alternate fish species. Croakers, lizard fish, barracuda, striped mullet, leather jacket, cod, hake, whiting, horse mackerel etc. have now been identified as other major species for surimi processing. The important criteria in the selection of fish for surimi are :

- low cast.
- white fleshed (non-oily).
- abundantly available round the year.
- good gelling ability.

During rigor the fish will become rigid and hence pose difficulties in handling and separation of meat using a meat/bone separator. Therefore, post-rigor fish is used in surimi production. Though freshness is an important criterion, fish stored in ice for 3-4 days is used in several processing establishments. However, fresh fish is more preferable as the blood and gut residues will be less in the tissues. Such fish might have undergone only a lower degree of autolysis compared to stored fish and hence will yield surimi of better gel.

Small pelagic fishes like sardine and mackerel are abundantly available in tropical waters; however, the surimi processed out of them is inferior in quality because of the high fat content, high myoglobin content and the presence of dark muscle. Some improvements can be brought about in fatty fish mince by washing it with sodium bicarbonate solution. This will remove most of the fat. Majority of the dark muscle can be removed using a pressure shower.

The mince for processing surimi can be made in two different ways. The fish can be filleted and the fillet passed through a meat/bone separator to produce the mince. The yield of mince will be lower in this process, but will produce surimi better in colour, gel forming ability and requiring fewer washes and no need of refining. The second process involves producing mince using a meat/bone separator from headed and gutted fish. This process gives higher yields than the other one.

In commercial practices three cycles of washing with 10 minutes contact with water: mince ratio of 3:1 or 4:1 are commonly adopted. The water temperature is usually maintained at 5-10°C to prevent protein denaturation.

12.3.2.2 *Washing*

Cyclic washing of mince with water is the most important process in surimi production. Washing is essential to remove

- water-soluble sarcoplasmic proteins which do not contribute to gel formation.
- the protease enzymes.
- blood and pigments.
- lipids.
- haem compounds responsible for lipid oxidation leading to protein denaturation.

Water quality is another important factor deciding the quality of surimi. Hard water containing ions of calcium/magnesium will cause textural degradation and iron/manganese will affect colour during storage. Since fresh mince has pH in the range 6.5 to neutrality, water used also should have a pH close to that of the mince to facilitate retention of its original water holding capacity. Salt is added at the level of 0.1-0.2 per cent in the last washes as this will aid in dewatering. Higher levels of salt will solubilise actin and myosin and affect the gel strength of surimi.

Though mince can be made using whole fish in the meat/bone separator, it is not considered desirable because it will have poor colour, have high oil content and protease enzymes from the guts which will tend to impair the gel forming ability.

12.3.2.3 *Dewatering*

After washing, the water content of the mince has to be reduced to about 85 per cent. Dewatering can be done by using a screw press or by centrifugation. Putting the mince in a cheese cloth-type fabric bag and hand pressing will be an ideal method, since the temperature of the mince will not rise during the pressing operation. For large scale operations dewatering can be done this way in batches or by centrifugation.

12.3.2.4 *Refining/Straining*

Refining is the process of removal of left out bones, skin and perhaps dark muscles from the washed mince. This operation can be performed either before or after dewatering. For proper straining the mince should have a moisture content in the range 80-90 per cent. Dehydration prior to straining will make it slow and difficult.

For refining, the mince is fed into a drum by a screw conveyor and pressed through the perforated drum. Residue such as bones, skin etc. is collected at the end of the drum.

12.3.2.5 *Addition of Cryoprotectants*

Cryoprotectants are added to the refined/dewatered surimi to enhance its quality and water-holding capacity. Uniform mixing of the cryoprotectants is achieved by using a silent cutter or similar machinery. Since fish meat is quite fragile precaution should be taken to avoid generation of excessive frictional heat. The cryoprotectants generally used are sucrose (4 per cent), sorbitol (4 per cent) and polyphosphate (0.3 per cent) either alone or in various combinations. Some other cryprotectants found suitable for addition to surimi are glucose, galactose, fructose and lactose. Surimi processed with the above cryoprotectants is called 'salt-free surimi'. Surimi to which salt is added is 'salted surimi'. Salted surimi may contain upto 2.5 per cent sodium chloride. Additions such as starch and flavouring agents also may be added.

The entire processing operations should be carried out in a cold atmosphere. At no time during the processing operations the temperature should be allowed to exceed 10°C. Higher temperatures will adversely affect the functional properties of the proteins.

Surimi is packed in trays, frozen in a plate freezer at -30 to -40° C and stored below -20° C until used. Maintenance of constant temperature in the storage is very important. Temperature fluctuations can affect the gel forming ability.

If the frozen surimi is stored at -10° C, the quality becomes rapidly impaired within one month and becomes unfit for further processing into surimi based products in three months.

12.3.3 Role of Additives

12.3.3.1 Sucrose

The role played by sucrose as a cryoprotectant of surimi is by preventing denaturtion of proteins, particularly of actomyosin, during frozen storage. Its effectiveness as a cryoprotectant is considerably increased when synergised with small amounts of polyphosphates. Sugars also increase the surface tension of water and consequently the amount of bound water.

12.3.3.2 Polyphosphates and Sodium chloride

Sodium chloride in lower concentrations enhances water uptake and swelling of surimi and products like sausage, ham etc. made using surimi. Phosphates prevent drip loss and improve the water holding capacity of the frozen surimi. When meat pieces are mixed in the presence of sodium chloride and polyphosphate, a sticky exudate forms at the surface of the meat and helps in its binding. Myofibrils swell in high concentrations of sodium chloride; however its quantity required can be considerably reduced in the presence of polyphosphates.

12.3.3.3 Starch

Starch can enhance the textural qualities of surimi gel by improving gelatinisation. Starch will facilitate greater water absorption. Firmness and cohesiveness of gel increase with viscosity and water holding capacity of the gelatinised starch.

12.3.4 Treatment of Wash Water

Surimi process involves use of enormous volume of water. This water can create pollution problems unless it is pre-treated before realese. One way to partially overcome this problem is to minimise the quantity of water used by employing wash water in a counter current re-cycle system. However, the water from the first wash that will contain the worst contaminants should not be re-cycled. The pollution problem can be minised by recovering the water-soluble proteins by flocculation, ultra filtration or some other suitable means. It has been demonstrated that the proteins recovered from wash water by micro- and ultra filtration can be included in the surimi without loss of gel strength.

12.3.5 Surimi from Small Pelagic Fishes

Production of surimi from small pelagic species is considered more difficult because of

- the higher lipid content.
- comparatively large amount of sacroplasmic proteins.
- higher content of dark muscle.

The important differences in the process employed for production of surimi from small pelagic fishes are the use of

- sodium bicarbonate solution for washing the mince.
- a superdecanter to separate fish oil and water from the meat.
- a pressure shower to mechanically remove the dark muscle, skin and subcutaneous fat and to collect the ordinary muscle.

Washing with sodium bicarbonate solution is very effective in removing the fat from the mince. This is followed by separting the oil and water from the meat using a superdecanter.

12.4 ONBOARD PROCESSING OF SURIMI

Sixty per cent of the Japanese surimi is processed on board factory vessels. However, in the U.S. high quality surimi is produced in shore-based plants using fish delivered in refrigerated sea water. The most important factor in the on board processing is assurance of consistent fish delivery at low price. Advantages of on board processing and shore-based processing are as follows.

- On board processing.
- proximity to resource.
- lower price for raw material.
- regular delivery of fish.
- ability to 'follow' the resource.
- easy and abundant supply of fresh water (desalination of seawater).
- Shore-based plants.
- unlimited water supply
- lower labour cost, the workers being hired from the locality.
- flexible use of the space in the plant.
- lower repair cost of equipment since the plant is more accessible.
- storage and transportation cost are less.

12.5 QUALITY ASSESSMENT

The main criteria employed in judging the quality of surimi are gel strength, colour and flavour. These are influenced by

- the species of fish used.
- freshness of the fish.
- processing method employed.
- moisture content.
- freezing and storage temperatures.
- handling and distribution methods.

In addition to the above, organoleptic tests, particularly folding test and teeth cutting test, are also performed to assess the quality of surimi.

12.5.1 Grading of Surimi

The Japanese system classifies surimi according to the properties which are important to kamaboko production, i.e., colour, moisture, gel strength and contaminants. It does not, however, grade a product that may score high in one parameter, e.g., whiteness, but less in another, e.g., gel strength.

The potential uses of surimi are far more diverse in the U.S., where it is being increasingly used as meat binders, protein drinks, snack foods etc. Hence, the grading system must focus on functionality.

12.5.2 Measurement of Gel Strength

For measurment of gel strength, a sample of surimi is made into a paste by grinding with 2.5 per cent salt and 30 per cent chilled water for 30 minutes. The paste is then filled into a PVC casing and allowed to set in a water bath maintained at 40°C for 20 minutes, followed by heating at 90°C for 20 minutes. 25-35 mm long samples are cut out for gel strength measurement.

A widely used method of gel strength measurement is the penetration method using a Rheometer. The most commonly used instrument has a plunger with a sphere of 5-10 mm diameter at the tip of a 10 cm long rod. The plunger is pressed on to the surface of the sample until it penetrates through the surface. The gel strength is expressed in g-cm as the product of the force required to break the sample and the depth before penetration.

12.5.3 Determination of Whiteness

Whiteness is usually measured using a whiteness meter and is expressed as per cent Lovibond after comparing with standard whiteness of 93 per cent Lovibond pure whiteness.

12.5.4 Organoleptic Tests

The organoleptic tests, folding test and teeth cutting test, have to be carried out by trained personnel. In spite of likely personal variations, these tests can give a quick assessment of the springiness of the product. The first step in the organoleptic tests is the preparation of the sample which is done in the same way as for gel strength measurement.

12.5.4.1 *Folding Test*

Five slices of the prepared sample are taken and are first folded in half. If no breakage occurs at this stage they are folded into quarters. Grading are given in Table 12.1.

Table 12.1 : Test Samples and their Grades

Condition of test samples when folded	Grade	Remarks
No breakage in any five samples when folded into quarters	AA	Very good
Slight tear in any one of the five samples when folded into quarters	A	Good
Slight tear in any of the five samples when folded in half	B	Fair
Breakage, but two pieces still connected, when folded in half	C	Poor
Complete breakage into two pieces when folded in half	D	Very poor

12.5.4.2 *Teeth Cutting Test*

The sample for testing is prepared as for the folding test. The springiness of the sample is felt by biting the slice between upper and lower incisors and the scoring is made on a hedonic scale (Table 12.2)

Table 12.2 : The Springiness and the Score

Description	Score
Extremely strong springiness	10
Very strong springiness	9
Strong springiness	8
Quite strong springiness	7
Acceptable springiness	6
Acceptable slight springiness	5
Weak springiness	4
Quite weak springiness	3
Very weak springiness	2
Mostly no springiness	1

12.5.4.3 Microbial Quality of Surimi

Only fresh fish or fish properly iced stored only for a very limited period is used for processing into surimi. Generally the fish is gutted and headed before converting into mince for making the surimi. The intestines and gills, which are the major areas where the bacteria harbour, are thus eliminted before processing. The product is free from the skin which is another area where the bacteria harbour. The deboning machine is a potential source for bacterial contamination; but the subsequent repeated washings and, the use of cryoprotectants and salt minimise this hazard. The chilled water used in processing and the overall process schedules carried out under cold temperatures may help development of psychrophilic bacteria. However, the surimi is subjected to heat treatment during product manufacture which will result in the elimination of the remaining organisms. In general, it is reasonable to assume that surimi quality is not significantly affected by bacterial growth.

12.5.5 Surimi-based Products

12.5.5.1 Fibreized Product

Fibreized products are the greatest in demand among the surimi based simulated shellfish products. The ingredients used in the formulation of fibreized products include surimi, salt, starch, egg white, shellfish meat, shellfish flavour, flavour enhancers and water. The mixture is ground to a paste using a silent cutter. It is then extruded as a thin sheet onto a conveyor belt and is heat treated using gas and steam for partial setting. After cooling, a strip cutter sub divides the sheet into strings which pass through a rope former. The rope is coloured and shaped. The final product is formed by steam cooking the coloured and shaped material.

Imitation crab leg, shrimp, lobster and scallop meat are the fibreized products high in demand.

12.5.5.2 Kneaded Products

Kamaboko, chikuwa, hampen, fish ham, sausage etc. made using surimi are sometimes referred to as kneaded products. The classification is principally based on the manufacturing process involved.

• KAMABOKO

Kamaboko is one of the oldest traditional paste fishery products of Japan. Originally kamaboko used to be made out of fresh fish. Presently it is mostly made out of surimi. Surimi is ground to a fine paste with other ingredients which will include salt, monosodium glutamate, sugar, starch, sweet 'sake', egg whie, eggs or polyphosphates and water. Grinding time varies from 30 to 40 minutes. The ground paste is fixed on a thin wooden plate made of pine or cedar in the shape of half cylinders or bread loaves. It is then steamed in open air for periods varying from 80-90 minutes for large size, 40-60 minutes for medium size and 20-30 minutes for small size. The maximum temperature attained at the centre of the paste is 75°C although the surface temperature attained is 90-95°C. Kamaboko can also be made by boiling instead of steaming.

Good kamaboko is quite white, and has a smooth appearance with elastic texture. It can be made to various recipes to suit the regional preferences, fish species available, additives and seasoning used etc.

• CHIKUWA

Chikuwa is made by moulding surimi ground with other ingredients round a brass tube or a piece of bamboo and then broiling. The tube or bamboo is rolled and moved inside the oven as broiling proceeds.

Chikuwa has a hollow cylindrical shape. Good chikuwa is white inside with brown surface.

• HAMPEN

Surimi is ground to a fine paste along with Japanese yam and other ingredients. Different from kamaboko (steamed) and chikuwa (broiled), hampen is processed by boiling. Good hampen has a soft spongy texture resulting from incorporation of air during rapid kneading.

• FISH HAM

A method of processing fish ham widely practised is as follows. Salt, sugar, chemical flavourings, spices and a smoke flavouring are added to the surimi made from Alaska pollack and the mixture is allowed to cure for a day or two in a cold atmosphere. The cured mass is mixed with additional quantity of surimi, starch and soy protein and ground to a fine paste. The paste is sealed in PVC casings and is heated in water at 85-90°C for 40-60 minutes and frozen after cooling. If the product is intended for marketing fresh, it will be steamed for four minutes at 120°C.

• FISH SAUSAGE

Fish sausage is identical to pork sausage. Surimi is mixed with salt, sugar, sodium glutamate and soy protein in a definite ratio in a silent cutter. At the end of mixing, lard or shortening, polyphosphate and smoke flavourings are added and the homogenised mass is stuffed into PVC casing using automatic screw stuffer. The casing tube is closed using metal rings. The tube is heated in hot water at 85-90°C for 40-60 minutes. After heating it is slowly cooled to avoid shrinking of the tube and then dried.

Suggested Reading

Clucas, I.J. and Ward, A.R. *Post-harvest Fisheries Development: A Guide to Handling, Preservation, Processing and Quality*. Chatham Maritime, Kent, ME4 4TB United Kingdom.

Gopakumar, K. (1997) *Tropical Fishery Products.* Oxford & IBH Publishing Co., New Delhi.

Granthan, G.J. (1981) *Minced Fish Technology : A review.* Fishereis Technical Paper No. 216FAO, Rome.

Min, T.S., Chin, N.M., Fujiwara, T., Kuang, H.K. and Hasegawa, H. (1987) *Handbook on the Processing of Frozen Surimi and Fish Jelly Products in Southeast Asia.* Marine Fisheries Department, SEAFDEC, Singapore.

Putro, S. (1990) *Processing of Surimi and Fish Jelly Products.* INFOFISH Technical Handbook No.2. Infofish, Kuala Lumpur.

Suzuki, T. (1981) *Fish and Krill Protein Processing Technology.* Applied Science Publishers Ltd., London.

Tanikawa, E. (1971) *Marine Products in Japan.* Koseisha-Koseikaku Company, Tokoyo.

13
BATTERED AND BREADED PRODUCTS

COATING seafood, poultry or vegetable products with a batter and/or breading before cooking is an established domestic as well as commercial practice. Of late, the emphasis of coated products has shifted from the home front to the restaurant and fast-food outlets.

Prominent among the coated products are the seafood products. Battered and breaded fish finger or fish stick has been one of the forerunners in this field and still command a sizeable share of the battered and breaded seafood trade. Such products, when frozen in a ready-to-cook form, offer a convenience of high consumer value and are called 'convenience foods'. Fish finger, fish portion and fish cake are the staple battered and breaded fish products. Breaded shrimp, scallop and oyster cater to the luxury market.

13.1 COATED PRODUCTS

A coated product, also called enrobed product, is one, where a food material is coated with another foodstuff. A coating will be referred to as the batter and/or breading adhering to a food product after cooking. A batter can be defined as a liquid mixture composed of water, flour, starch and seasonings into which food products are dipped prior to cooking. The breading is normally a bread-based crumb, but other coatings like small potato chips or puffed grain such as rice also are popular.

13.1.1 Functions of Coating

Batters and breadings have several functions to perform as food coatings :

- enhance the appearance of food products.
- enhance the taste characteristics by providing food. products with more crispy texture.
- improve the nutritional value of the product.
- provide a more desirable colour.
- act as moisture barrier and minimise moisture loss during frozen storage and microwave reheating.
- act as food sealant by preventing natural juices from flowing out and seal in the flavour.

13.1.2 Coating Ingredients

The unique characteristics and functionality of the coating are the sum total of the characteristics of the ingredients going into their making. Coating ingredients are grouped under the following categories :

- polysaccharides - wheat, corn flour, starch, farinaceous materials, modified derivatives of cellulose, and gums.
- proteins - various milk powders and milk protein fractions, egg albumin, cereal flours, seed proteins.
- fats and hydrogenated oil.
- seasonings - sugar, salt, pepper and other spices or spice extractives.
- water.

13.2 FUNCTIONS OF COATING INGREDIENTS

13.2.1 Flour

Flour, by usual definition, specifically refers to finely ground endosperm of wheat. However, in battering and breading process, flour refers to finely ground starchy material from several sources including corn, rice, soy or barley. Functionality of flour in batter and breading systems depends mainly on the major constituents of flour, viz., starch and protein.

13.2.1.1 Protein

The functionality of the wheat flour in batter is contributed by the development of gluten proteins. Because of their higher protein content, hard wheat flour require more water in the mix than soft wheat flour to yield a batter of comparable viscosity. This results from the efficient water binding capacity of the gluten protein.

Viscosity is an essential factor which determines the amount of coating 'pickup.' Pickup refers to the incremental added weight from the coating operation expressed as percentage of the total weight of the finished product. Batter viscosity depends on the ratio of flour mix to water and the temperature of the mixture. Many ingredients in a batter are insoluble in water and hence a more viscous batter will bring about the suspension of such ingredients preventing undesirable settling. Flour proteins help maintain such ingredients uniformly dispersed. In tempura batters, gluten proteins help in gas retention during leavening.

13.2.1.2 Starch

Starch in the flour is made up of linear and branched polymers of glucose (amylase and amylopectin). The ratio of the polymers significantly influences the functionality of the starch. Starch is made up of granules of varying sizes. In intact cereal grain the starch granules are embedded in a protein matrix. The degree of binding between protein and starch and the process used to separate them influence the particle size distribution of the resulting flour. The particle size and degree of physical damage of starch granules affect the functional properties of flour. Damaged starch is the granules mechanically altered during milling. Damaged starch granules greatly increase the water absorption capacity over undamaged granules. Therefore, with flour having higher proportion of damaged starch the amount of water for a given batter viscosity will be more than the normal.

Starch undergoes gelatinisation on heating in presence of water. In the original granule structure a considerable portion of the starch exists in a crystalline form, which at moderate temperatures is impervious to water. On heating, these areas are broken down and become exposed to water resulting in swelling of the sample. Further, some material from within the starch granule is exuded into the solution and further increases its viscosity. On cooling, the gelatinised starch molecules re-aggregate and form a gel that gradually increases in firmness with time and lower temperature. The process is called retrogradation. The rate and degree of retrogradtion is more rapid for linear amylose than for branched amylopectein. Gelatinised starch along with the flour proteins will form the structure of the final cooked product coating. For an even base coating the structure in the batter must be evenly distributed around the product to ensure the formation of a uniform starch gel that completely jackets the product.

The temperature of gelatinisation of starch varies depending on the source. In a batter where water is limited, complete gelatinisation can occur only at high temperatures and the swelling process will be limited. Thus it has to be ensured that a batter formulation provides a viscous medium and also provides enough water to effect sufficient degree of gelatinisation of starch. However, it must be remembered that with increasing water content the batter will become thinner with reduced pickup. This is compensated by adding various gums to the batter.

13.2.1.3 Non-wheat Starch

The sources of other starches used in the batter are corn, rice, soy and barley. Cornstarch is a source of natural yellow carotene pigment and hence it can supplement browning agents like reducing sugars and milk powder to impart a golden brown colour to the coatings. Cornstarch is used also as a carrier of spices as also to mask the starchy units from other ingredients. Cornstarch also helps to improve the crispiness of the coatings. Particularly in the case of tempura batters, corn helps to reduce the brittleness of the gluten protein. It is also possible to make flour-water slurries of wide ranging viscosity through binding of corn milling fractions. Rice, soy or barley can be added to batters and breadings for increased adhesion and water holding capacity. Such additional water will be available for modification of viscosity at room temperature and starch gelatinisation during heating.

13.2.1.4 Modified Starches

Starches are often modified in various ways to provide specific functional effects. The simplest and most common modification is pregelatinisation. In this process starch is heated in water until gelatinised, and then dried to a powder. Extensive modification of starch includes changes in the degree of branching (variations in the amylose and amylopectin content), average chain length and the extent of cross-linking. These extensively modified starches are known to increase the adhesion of breading with the product.

Addition of pregelatinised starches to batter/breading mixes will provide similar functionality as obtained by using flour with more highly damaged starch.

13.2.2 Leavening Agents

Sodium carbonate and one of a number of leavening acids/salts are used to produce carbon dioxide, the leavening gas, in puff or tempura batter. The release of carbondioxide into the batter is controllable by selection of an appropriate leavening acid to react at a given temperature. Some batters use a mixture of acids, one of which will produce gas at room temperature and the other at high temperature to provide leavening effect during frying.

The addition of acid is based on the neutralising value and the rate of reaction of the leavening acids with sodium bicarbonate. Neutralising value is defined as the parts of leavening acid required to react completely with 100 parts of sodium bicarbonate. The leavening acids/salts in common use are tartaric acid, potassium hydrogen tartrate, monocalcium phosphate monohydrate, monocalcium phosphate anhydrous, sodium acid phosphate, glucose-d-lactate, sodium aluminium phosphate, dicalcium phosphate dihydrate and sodium aluminium sulphate.

13.2.3 Egg

Egg contains albumin, a heat-coagulable protein that is useful in binding both breading and batter to the product and to itself. The yolk protein contains lecithin, which being an emulsifier will contribute to batter stability. Addition of egg to a batter will tend to darken the final product. This will also add characteristic 'eggy' note to its flavour.

13.2.4 Milk and Whey

Added as liquids or dry powders, milk and whey provide lactose, which is a reducing sugar, involved in browning reactions. However, they also contribute proteins which provide structure.

13.2.5 Salt and Sugar

Salt is added to many batter/breading formulae primarily as a flavouring agent. In limited water systems salt will compete with flour proteins for water and tend to slow the rate of protein hydration. Occasionally added for a sweet flavour, sugar also will compete for water, but as with salt this is not a problem in batter/breading applications.

13.2.6 Spices

Many spices, particularly pepper in combination with small amounts of others, are frequently included at a 3-5 per cent level of batter/breading formulae. Some spice like paprika also imparts some colour in addition to flavouring. Spices are not known to interfere with the functionality of the batter/breading ingredients.

13.2.7 Gum

Many of the hydrocolloid substances known as gums are used as ingredients in batters and breadings. Gums control the viscosity and water holding capacity. Some gums have the ability to participate in a gel or film formation in conjunction with other ingredients. This strengthens the coating and may provide greater flexibility and increased resistance to handling.

Gums are effective at less than two per cent of the formula and often in the 0.5 per cent range. Mixture of gums is also used in batter formulations for specific results. Because of the lower levels in the formula, gums do not dilute the protein of the base flour. Among the natural gums, xanthan has been of particular use in viscosity control.

13.2.8 Shortenings and Oil

Shortenings and oil contribute to the overall flavour and mouth feel. They also tenderise the coatings due to lubricity and interaction with other ingredients. They may also function as moisture barriers, emulsifiers and antistaling agents. Breadings are often encapsulated with fat to produce a 'fried-like' flavour to oven or microwave reconstituted coatings.

13.2.9 Prepared Breadings

Prepared breadings are materials which are applied to battered food products to enhance their appearance and/or organoleptic qualities. Bread crumbs, corn flakes and the fragile bread crumbs produced as 'Oriental crumbs' are included under this group. All these are cereal based, pre-baked products, which can be expected to maintain their integrity in the batter through the cooking process.

Size range of particles, colour, flavour and compatibility with the existing processing system in use are the major considerations in the use of prepared breadings. In products like fish finger the particles should be smooth. It can be coarser where a rough look is needed. Prepared breadings are applied over a wet batter and are expected to adhere to it, which is largely governed by the viscosity of the batter. Coarser breadings will need a thick batter layer to hold them.

13.3 BATTERS

Batters can be broadly classified into two groups - the conventional or adhesive batter and, the 'puff' or tempura batter.

13.3.1 Adhesive Batter

An adhesive batter coating is always used with a supplemental breading or bread crumb. The primary purpose of adhesive batter is to increase the crumb adhesion by acting as an interface between the food and the subsequent coating.

Uniformity and thickness of the coating are important aspects influencing the acceptability of the finished product. The adhesion batter achieves this by means of viscosity development. The adhesive properties and the percentage pickup are the two major considerations in the formulation of adhesion batter. The formulation and viscosity of the batter determine the amount of coating pickup. The pickup by the product will be more if the viscosity of the batter is higher. Consistent batter viscosity is essential to produce uniformly coated products. Batter viscosity is dependent on the ratio of the flour to water and the temperature of mixing.

The simplest adhesive batter is wheat flour mixed with water. A typical ratio of batter mix to water is 1:2. High proportion of water in the batter will require more time to freeze the batter and so to 'fix' the crumbs on the surface of the base product. For a 'quick set' batter it is necessary to use, in addition to wheat flour, some quantity of corn flour and/or starch and gum. The amount of water used will be correspondingly reduced.

The batter should be kept cool to avoid the growth of microorganisms. If the temperature goes up the viscosity will fall and the batter will not stick to the product. However, if the batter temperature is too low, and the fish to be coated is frozen, the batter may freeze on the conveyor belt and may leave uncoated areas on the product, the frozen batter being dragged off the fish by the moving belt. Tempering the fish prior to battering will overcome this problem.

13.3.2 Tempura Batter

Where a batter is required to provide an aerated crisp coating with or without the application of any other coating, a combination of wheat and corn flour are used along with a chemical raising agent. The most important difference between conventional and tempura batters is that the tempuras are used at very high viscosity levels and always contain raising agents. The batter mix is supplied in a powder form and is reconstituted with water to give the desired viscosity. The final texture is achieved on frying the coated product in oil at 180-220°C.

Mixing the batter is achieved by agitation. If too much air happens to be incorporated in the batter, the small air bubbles will agglomerate and coalesce into a large bubble on the surface of the fish. This may be blown off when the coated product passes under an air knife leaving a void in its position. During subsequent frying, the fish flesh in such areas will come into direct contact with hot oil and the water in the flesh will flash off as steam and will blow off the batter surrounding the void.

Because of the above problem the tempura batters are not generally pumped in the enrober. Submersion is used rather than overflow batter application when applying tempura batter. The enrober is designed to carry the fish between two conveyor belts through a bath of batter, ensuring that the fish is completely immersed. The gap between the upper and lower belts is made adjustable to accommodate products of different thickness.

Batter temperature, viscosity and mixing are critical, not only for pickup, but also to prevent loss of gas before coating.

13.4 BREADINGS

The secondary coating is referred to as 'breading' though it may not always be derived from bread. The original crumbs were ground dry bread and even today it is the major secondary coating. However, a wide variety of breading materials are available now in different sizes and colours. They can be used alone or combined with various crumbs, flours, starches and flavouring materials such as herbs, spices and seeds. Breadings are thermally processed cereal based products though non-cereal products like potato are also used to provide different textures and appearance to the end product. The particle size of the breading is important in terms of appearance, texture and pickup.

13.4.1 Breading Characteristics

The functional characteristics of breadings depend on the specific physical and chemical attributes built into the particular breading. The important considerations are the particle size, area to volume relationship, browning rate, moisture absorption, colour and texture, and oil absorption.

13.4.1.1 Mesh Size

Breadings are made in different particle sizes - finer to coarser. The proportion of the various fractions governs the final appearance of the product. The proportion of finer to coarser particles also affect the rate of absorption of moisture from the batter or from the fish itself on the processing line. The finer particles rapidly absorb moisture in few seconds from any batter. The larger particles provide visual appeal and textural impact. However, use of a very coarse breading on a food product with small surface area may result in the coating falling off on handling or transportation.

13.4.1.2 Area to Volume Relationship

Some foods having natural shapes have a particular area to volume relationship. Some are sliced or formed into various shapes and the ratio of their area to volume can be adjusted. A high area to volume relationship permits a good coverage to be applied. In some cases where the area to volume relationship is unfavourable, like a product cuboid or approaching a cube in shape, it becomes difficult not only to apply the coatings, but also to ensure pickup at economical levels.

13.4.1.3 Browning Rate

Some amount of browning is identified with the product quality in coated products. Browning rates of breadings depend largely on the proportion of reducing sugars used in their manufacture. Corn syrup solids, whey powder, milk powder, lactose etc. are the sugar sources used. Browning takes place during frying the coated products in oil. Fast browning rates will permit high processing speeds as also the choice of low frying temperatures.

13.4.1.4 Moisture Absorption

The rate of absorption of moisture by breading is a function of its particle size, porosity and gelation. A breading of smaller average mesh will have higher area and hence the coated products will move rapidly to a point where it can be handled conveniently. However, larger granules improve the appearance as also the texture of the product. The larger granules will protrude from the surface of the coating and the protruding peaks provide colour highlights as well.

A porous breading will absorb and release moisture and frying oil more quickly than a dense granule. It is porosity, together with mesh size, that determines the rate of absorption and largely the texture of the coating. These two factors are modified by the intensity and type of heat treatment that the breading undergoes during its manufacture. In order to balance the factors of appearance, texture and drying rate, various heat treatments are employed for a wide range of breadings.

13.4.1.5 Colour and Texture

The fried colour of the breading is contributed not only by the browning of reducing sugars, but also by the added colours. Permitted food colours, paprika extracts, synthesised carotenes or tomato pigments can be used.

Porosity of the breading also plays a role in the colour formation. Granules with an open structure can absorb frying oil and hence heat more rapidly than dense granules. Therefore, for the same level of reducing sugar content, porous granules change colour more rapidly than dense granules of the same composition.

13.4.1.6 Oil Absorption

Absorption of oil and effective rate of heat transfer is higher in porous than in dense granules. When a prefried product is reheated the absorption of the oil and the exchange of this oil for moisture during frying stage have an important advantage in texture development.

It can be understood from the foregoing that all the major characteristics of breadings interact to produce a wide ranging textural and colour preferences of the consumers of the breaded products.

13.4.2 Breading Types

A wide variety of breading materials are available in different sizes and colours. They can be used alone or in combination with other types of crumbs, flavours, starches and flavouring materials.

13.4.2.1 Reclaimed Bread Crumbs

These are prepared from ordinary bread, which has gone stale, by drying and grinding. Sometimes the drying process is carried out deliberately at a high temperature to give an effect of toasting and to reduce the bacterial load.

13.4.2.2 Industrial Crumbs

This is factory-baked in large volumes and is widely used as crumb coatings in fish fingers/sticks and other products. The production process is similar to conventional bread making. It uses a raising agent, uses lower quantity of water and is baked as a continuous loaf before grinding. Natural colouring agents like paprika or turmeric are widely used during the manufacturing process to impart an appetising appearance. Industrial crumbs tend to have a harder texture and higher density than the reclaimed bread crumbs and is manufactured to much more exacting standards.

During baking, a crust develops on the surface of the loaf. This is darker and harder than the rest of the crumbs. Although efforts are made to minimise this, the presence of crust particles is a feature of this type crumb. The crumb will brown quickly when fried.

13.4.2.3 Japanese Style Crumbs

The Japanese crumbs, also called oriental or Panko crumb has a characteristic flake-like elongated structure which has excellent visual appeal and provides a unique surface texture when fried. It has an open and porous texture which imparts a light tender crispiness. Because of its lightness, it is possible to produce the crumb in large sizes without the sensation of hard particles.

Japanese style crumbs are baked using an electrical induction heating process. The bread is baked in less than one half the time taken for conventional baking and results in a loaf that is crust-free and of low density. The loaves are cooled, shredded through specially designed mills and dried to a low final moisture level.

13.4.2.4 *Extruded Crumbs*

Extruded crumbs are produced by a continuous process where high starch ingredients are cooked under pressure. When the pressure is suddenly released, the moisture expands rapidly as steam and the extrudate expands. In the extrusion cooking process the heated dough exits from the extruder die as a fully cooked glassy material that immediately dries to a low moisture. At the high exit temperature the low initial moisture is quickly flashed off and, in effect, there is no drying system required.

Because of its lighter density the extruded crumbs have a tendency to float in oil, potentially leading to contaminating black spots in the fryer and rapid deterioration of oil quality.

13.4.2.5 *Cracker Meal*

The flour is made into dough with water and is rolled into a thin sheet and baked. The amount of effective cooking is adjusted not only by the baking time and temperature, but also by the dough thickness and the water to solid ratio in the dough itself. The baked sheet is then crumbled through a granulating mill or a slow speed grinder. It is then dried to final moisture content of approximately eight per cent. The dried coarse particles are roller milled, sifted and blended as required to arrive at the appropriate granulation specifications.

Cracker meal is principally used as coating for products which are deep fat fried for long periods and is widely used on fish products. Unless other ingredients are added, cracker meal does not develop colour as quickly as other breadings.

13.5 APPLICATION OF COATING MATERIALS

Seafood specialties such as fish portions (raw and precooked), shrimp, fish fingers, scallops and fillets are the principal seafood products which are battered and/or breaded. Most of these items are processed frozen, although some may be processed fresh. The important steps involved in the production of a coated fish product are discussed below.

13.5.1 Predusting

Generally, products like fish portions or fingers undergo a predusting step before battering. Products entering the coating line in a frozen condition quickly pick up a fine layer of ice through condensation from the atmosphere. This forms a barrier between the product and the batter resulting in poor adhesion. Predusting is done primarily to create a surface more conducive to the physical adhesion of a wet batter. Predusts provide a highly absorbent layer which forms a bridge between the raw material and the batter mix. Predusts also provide a rough surface which helps the batter to coat the product evenly and obtain the desired pickup.

The predust is usually composed of a cereal flour or flour mixture. It may also contain spices and seasonings for both functional and flavouring purposes. Salt is also used to slightly melt the ice glaze covering the surface of some fish portions. The additional moisture generated can hydrate the predust and create a surface more conducive for the wet batter adhesion.

13.5.2 Application of the Batter

Conventional batters are of low to medium viscosity and hence can be applied with total submersion or overflow batter applicators. Low viscosity batters are normally applied in an overflow configuration. Medium viscosity batters may require a total submersion system depending on the product requirements.

The predusted product is conveyed to the batter applicator and transferred to the next conveyor which will draw it through the batter. The fish portion is totally submersed in the batter as it is drawn through it. Other applicators may use a pour-on application in addition to the submersion method. Irregular shaped products should be placed on the line with any concave surface upward to prevent air pockets from inhibiting batter pickup.

Line speed is a very critical factor affecting batter pickup. An excessively fast line speed will reduce the batter pickup. The battering may become incomplete. There may not be enough time for the excess batter to drip off, and this excess batter will be blown off during prefrying. The blown off batter will get deposited in the fryer. Too low a line speed also can result in excessive batter adherence. The batter weight in the prefried products is adjusted to be equivalent to fish flesh weight in most seafood products.

Excess batter, if carried over to the breading section, will cause formation of lumps and can cause blockages in the breading machine. This will also cause formation of shoulders and tails on the edges of the product and contaminate subsequent breading application. Therefore, to overcome these problems the excess batter is removed after coating by blowing air over the product. The position of the air blower should be as close to the product as possible to control the airflow across the product.

Carry over from the predusting operation also is critical. Where predust is carried over, the viscosity of subsequent batter will increase leading to an increase in pickup.

13.5.3 Application of Breadings

There are many types of breading applicators available and the appropriate machine depends on the ingredients used. The speed of the breading machine is so adjusted to closely match the belt speed of the batter applicator.

For soft products the crumb depth should be maintained as thin as possible to avoid product damage when leaving the breading machine; however, frozen or hard products should have a deep bed of crumb.

Pressure rollers are used to apply sufficient force to press crumbs onto the battered product. But the pressure should not be high to distort the product shape or push the product through the crumb bed causing marks on the underside when the product may contact the breading conveyor.

Flour breadings have a tendency to compact and build up on the conveyors. They also tend to bridge and cake causing uneven flow through the breading machine which can result in inconsistent product quality. Due to their fine particle size, flour breadings tend to contaminate the frying oil with a residue so fine that it cannot be removed by normal filter systems.

Japanese style crumbs with their low bulk density and larger granule sizes make the crumb pickup difficult by the normal batter systems. Special batter formulations, sometimes containing raising agents, may have to be used at medium viscosities for a desired level of pickup of crumbs.

Specially designed breading machines are used to apply uniform particle size distribution or granulation to both the top and bottom of the product with minimum crumb breakdown. The machine used move the crumb by means of augurs. When two augurs are feeding crumb horizontally and then vertically, the augur speeds are adjusted to ensure that the horizontal augur feeding the vertical augur is operating at a lower setting. This avoids the crumb being crushed at the outlet end of the horizontal augur. If they are not synchronised, fines will result. When fines are produced in the machine, they tend to migrate to the edge of the conveyor belts resulting in non-uniform products across the belt. Adding oil at a level up to five per cent during the manufacture of crumb helps to reduce fines produced on the machine.

Air blowers are used to remove excess crumb from the fish after enrobing. Excess crumb carried into the fryer can cause unsightly black specks on the product. Filters are used to remove small particles from the oil to prevent this phenomenon.

13.5.4 Prefrying

Prefrying is another very important step in the battering process and serves the following purposes :

- sets the batter coating on the fish portion so that it can be further processed by freezing.
- develops the product colour.
- forms a characteristic crust typical of fried foods.
- provides the product a fried (oily) appearance which also inhibits freeze dehydration and contribute to taste.

This process is called prefrying because the final product frying for consumption is completed at the consumer's end. Before consumption the product is fried for about four minutes depending on the thickness and size of the product.

Both temperature of oil and duration of frying are critical to the development of the desired coating. Normally frying is carried out at a temperature of 180-190°C for about 30 seconds. Lower temperature and time will result in an undercooked coating. The reverse will result in a dark, overcooked coating.

In the prefrying process a wet coated product is entering hot oil. This produces an abrasive influence on the batter whereby any excess batter is forced away from the product and will get fried as separate entity. Such fried pieces are called 'tags', 'crumbs' or 'crunchies'. Some pieces may fall to the fryer bottom and clog the outlets. Some pieces may 'cake' onto the wire conveyer belt in the fryer and adhere to other fish portions. When such portions leave the fryer a part of fried coating may be pulled off leaving an unpleasant appearance. Some pieces may float in oil and pass out of the fryer with the prefried portions.

13.6 FREEZING

The temperature of the coating of the fish portions as they leave the frying oil will be equal to that of the oil. However, the product will still be frozen in the centre but the surface flesh may be partially thawed. Therefore, the fish portions coming out of the fryer are first cooled. This is accomplished by blowing ambient air across the surface of the portions. This allows the coating temperature to drop while allowing the batter coating to recover from the frying shock. During this period the coating stabilises itself, forms a coating which is more resistant to physical abuse and so the uncooked batter beneath the coating's crusted surface gets stiffened.

The prefried fish portions are generally frozen in two steps. These are initially individually quick frozen using liquid nitrogen or carbondioxide. After the initial quick freezing the portions are frozen in a slower manner using a mechanical freezer or even liquid nitrogen or carbondioxide. Freezing is continued until the internal temperature attained is around -12 to -15°C.

13.7 PACKAGING AND STORAGE

Most fish fingers, sticks or portions are individually packed into small boxes on a weight basis or by a specified number of pieces. They are also packaged in bulk form. When individually packed, layers of fish portions are separated by a waxed paper to prevent further product damage. This box is labeled and overwrapped with a polythene film to prevent moisture loss and freezer burn during storage. The packed boxes are stored at around - 20°C until shipped.

Proper packaging and maintenance of proper frozen storage temperature are two factors significantly affecting the quality of coated fish products during storage. On opening a stored box, if the box surface and fish portions have frost covering them it can be assumed that the product has undergone at least one freeze-thaw cycle. Such products may appear dark on reconstitution in a fryer. Poor package may also cause freezer burn in the product. A product with freezer burn may have the reconstituted coating flat and dull.

Suggested Reading

Current Manufacture's Literature. (1989) Stein, Inc., Sandusky, OH.

Darley, K., Dyson, D. and Grimshaw, D. (1982) *Production of Oriental Style Breading Crumbs*. US Patent 4,423,078.

Davis, P. (1977) Breaded seafood production in Australia. *Food Technology in Australia*. 29: 443.

John, H. (1956) Raw, breaded or precooked seafoods. In : *Preparation, Freezing, and Cold Storage of Fish, Shellfish and Precooked Fishery Products*. Fishery leaflet 430. US Fish and Wildlife Service, Washington, D.C.

Kulp, K. and Loewe, R. (eds) (1992) *Batters and Breadings in Food Processing*. American Association of Cereal Chemists, Inc., Minnesota, USA.

Manufacture's literature. (1989). Koppens Industries Inc., Stone Mountain, GA.

Quigg, J.R. (1980). Increased consumption of varied seafoods demands new batters and breadings. *Quick Frozen Foods*, 44 : 113.

Suderman, D.R., and Cunningham, F.E. (eds) (1983) *Batter and Breading Technology* The AVI Publishing Co., Westport, Connecticut, USA.

14
FREEZE-DRYING

FREEZE-DRYING, as the name indicates, is a process of dehydration, i.e., removal of water from the food material, while it is in the frozen condition. This is accomplished by subjecting the frozen material to a very high vacuum when water in its solid state, i.e., ice, sublimes into vapour without passing through the liquid stage. The principal difference between freeze-drying and hot air drying is that the latter involves removal of water from its liquid phase, i.e., it is a liquid phase drying and does not involve the use of

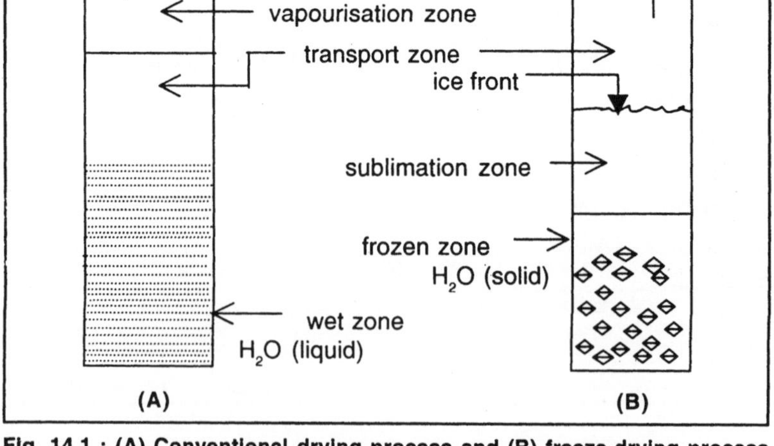

Fig. 14.1 : (A) Conventional drying process and (B) freeze-drying process

complicated equipment. Freeze-drying, on the other hand, is solid phase drying where the zone of dehydration in the food material is held under sub-zero temperatures during the process. Freeze-drying is also capital intensive as it employs sophisticated and expensive equipment. Conventional liquid phase drying and freeze-drying processes are represented in Fig. 14.1.

14.1 PRESSURE -TEMPERATURE RELATIONSHIP AND BOILING POINT OF WATER

Water boils at 100°C when the atmospheric pressure is 1 bar. If water is heated in a sealed container in which a vacuum is drawn, it will boil at progressively lower temperatures corresponding to the decrease in pressure; however, higher will be the latent heat of evaporation. This is illustrated in Table 14.1.

Table 14.1 : Pressure-Temperature Relationship of Boiling Water

Pressure (bar)	Temp. of boiling °C	Latent heat of evaporation kJ/kg
1	100	2257
0.5	81	2305
0.2	60	2358
0.1	46	2392
0.01	7	2485

Water can boil at 0°C, i.e. the temperature of ice/ice melt water, if the pressure is further reduced to 0.006 bar (Table 14.2). The latent heat of evaporation needed is 2500 kJ/kg; but the water would take some heat (334 kJ/kg) from itself causing the remainder to freeze and leaving as ice at 0°C. If the pressure is maintained at this level, ice will sublime, i.e., get converted straight into the vapour phase. If the pressure is reduced further, ice will sublime at temperatures below 0°C.

Table 14.2 : Pressure-Temperature Relationship in Sublimation of Ice.

Pressure (bar)	Temperature °C	Latent heat of evaporation kJ/kg
0.0060	0	2834
0.0026	–10	2836
0.0010	–20	2837
0.0004	–30	2838
0.0001	–40	2838

14.1.1 Triple Point of Water

Under a set of environmental conditions water can exist in equilibrium with its other two states, i.e., ice and vapour. This is referred to as the triple point of water. This condition is obtainable when the temperature of the water is 0°C and the pressure is 0.006 bar. At a pressure below this ice will sublime and can be removed from the system using a vacuum pump. This forms the basis of freeze-drying of foods. (Fig 14.2)

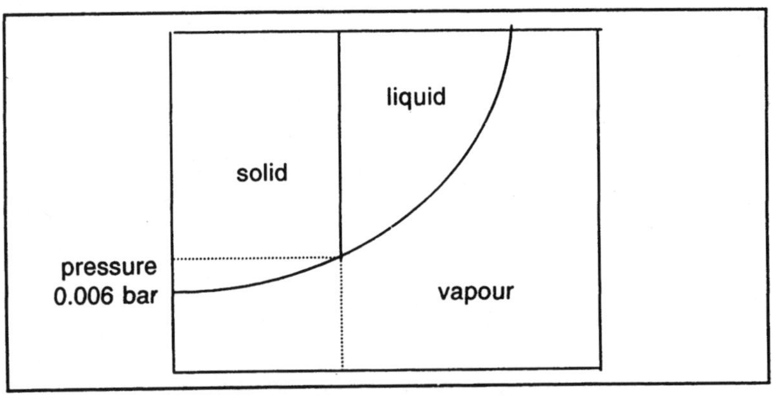

Fig. 14.2 : Triple point of water

14.2 FREEZE-DRYING

As stated above, water boils at 0°C when the pressure is 0.006 bar. The latent heat of evaporation taken from the water causes the remaining water to freeze. Ice sublimes below 0.006 bar pressure causing a further fall in the temperature of the ice because of the latent heat of evaporation taken from it. Sublimation of ice will continue until a stage is reached when further sublimation will cease unless some heat is supplied to the material from external source. To continue the process of drying, heat is supplied to the food material from an external source sufficient to continue the process without causing meltage of ice.

14.2.1 Factors Affecting Heat Transfer in Freeze-drying

In the classical freeze-drying process heat is supplied to the food by circulating heated fluids through shelves in the drying cabinet. Heat transfer takes place by conduction to the lower sides of the tray carrying food and by radiation from the shelves above to the upper surface of the food. Even if the temperature of the heating medium is maintained comparatively high, the amount of radiant heat transferred is rather very little. In addition, large pieces of fish show a tendency to curl away from the supporting tray. This appreciably reduces the area of contact of the fish with the tray, eventually resulting in very little conducted heat being transferred to the fish body. Even when initially there is a perfect contact of the fish surface with the tray it need not remain so throughout the drying cycle. The ice droplets escaping from the lower surface at a high rate may support the fish virtually on a cushion of high velocity vapour which will reduce the contact area.

14.2.2 Improvements in Heat Transfer Techniques - Accelerated Freeze-drying (AFD)

Although it is often faster than many other conventional methods of drying, freeze-drying is a comparatively expensive process. In order to reduce the production cost and increase the output, several improvements have been brought about to accelerate the process. The important physical problem in freeze-drying is improving the heat transfer to the product during drying while still maintaining an adequate vapour escape path.

In the beginning of the freeze-drying process ice is at or near the surface. At this stage maximum benefit could be gained from a high rate of heat transfer. It is also essential to permit ready escape of sublimed ice. High rate of heat transfer with simultaneous ready escape of sublimed ice can be achieved by interposing between the fish and the heating surface sheets of expanded metal, normally of aluminium.

Fish for freeze-drying is, therefore, arranged in a single layer between expanded metal mesh held in the tray. The tray is further covered with an aluminium or stainless steel sheet (Fig. 14.3). The trays thus prepared are placed between heating plates in the drying chamber (Fig. 14.4).

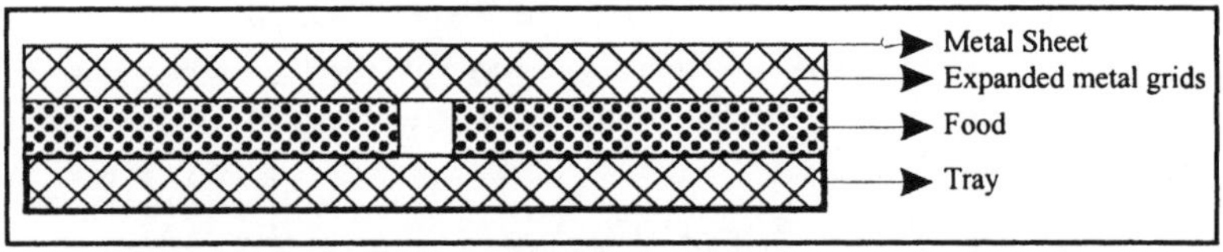

Fig. 14.3 Food and expanded metal arrangement for AFD

When the chamber pressure is brought down to the required level, often 0.001 bar, and the material is at the sublimation temperature, the fluids in the hollow plates are heated to a temperature between 60 and 100°C. The heat is conducted rapidly through the metal coverings to the fish surface causing rapid sublimation of ice. The vapours escape readily through the channels formed by the inesh into the chamber and is removed using the vacuum pump. If the fish is in direct contact with a continuous metal surface the vapour cannot escape readily causing the pressure at the fish surface to increase thus resulting in melting of ice. This contingency is obviated by maintaining the fish between expanded metal mesh.

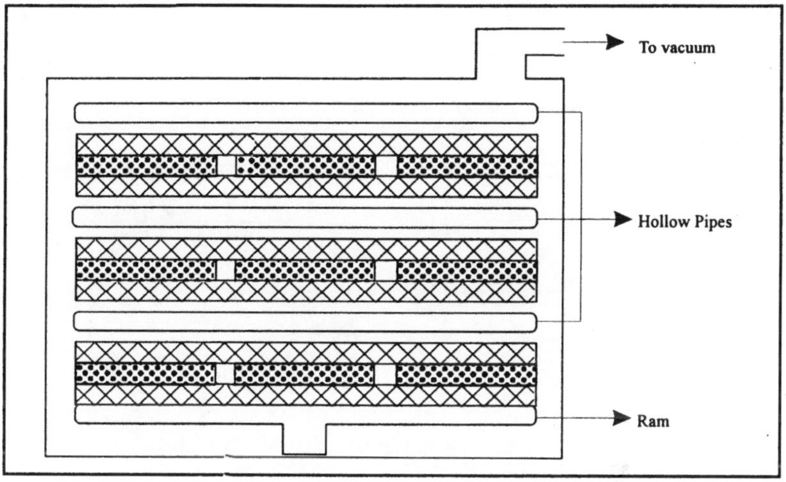

Fig. 14.4 Arrangement of trays in the chamber of AFD

As the sublimation front recedes, some pressure is applied to the plates using a hydraulic ram (see Fig. 14.4) moving the expanded metal sheet to penetrate slightly into the fish so as to provide more direct heat conduction to the sublimation front and not beyond that. The extent of pressure to be applied is predetermined in trial experiments and is around 0.5 kg/cm^2 for most foods. Simultaneously the temperature of heating is reduced. This is essential to prevent overheating of the already dried zone, because after sublimation the surface temperature of the fish will be the same as that of the heating surface.

Employing the above technique a very high rate of heat transfer is possible in the initial stages, as there is little resistance to the escape of vapour. The temperature of ice in the fish rises until the related vapour pressure is in equilibrium with the cabinet pressure and thereafter all the heat transferred to ice is absorbed as latent heat of sublimation. This method of controlled freeze drying providing contact of the heating surface with the material being dried from top and bottom with adequate provisions for speedy and effective escape of vapour is called accelerated freeze-drying (AFD). A set up for freeze-drying is schematically presented in Fig. 14.5.

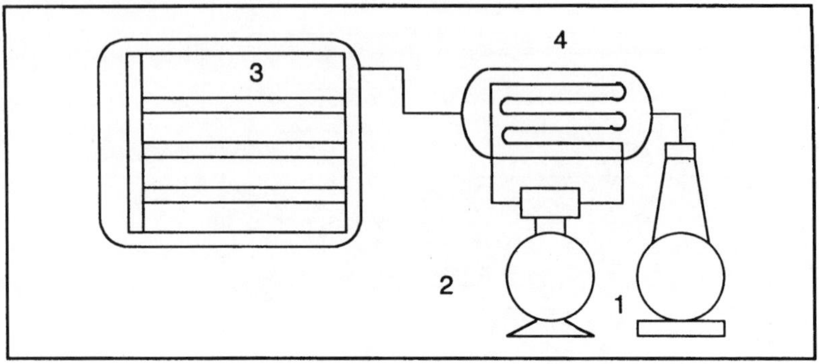

(1) vacuum pump; (2) refrigeration unit; (3) drying chamber; (4) condensation unit

Fig 14.5 Set up for freeze-drying

14.3 ACCELERATED FREEZE-DRYING PROCESS

The raw material is prepared in the same way as for dehydration prior to freezing. In the case of shrimp, it is size graded, peeled, deveined and cooked or, in some cases, cooked and peeled. Fish may be filleted, skinned and then frozen. It is very important that the size of the material going for drying should be uniform. The prepared material is arranged in trays in a single layer for freezing. Accelerated freeze-drying process involves the following steps :

- fish is placed in a single layer between expanded metal held in a tray. The tray is then covered with a thin aluminium or stainless steel sheet.
- the trays with the fish are introduced into the drying chamber and held between the hollow heating plates. The chamber is then closed air tight.
- the refrigeration and the vacuum systems are put on. The pressure level should be brought to 0.001 bar or lower (preferably 0.006 bar). (The fish becomes cool and ultimately frozen due to evaporative cooling taking the latent heat of evaporation from the fish body)

- once the fish attains the sublimation temperature which corresponds with the pressure, heat is supplied to provide the latent heat of sublimation. The frozen water sublimes and escapes through the mesh into the chamber from where it is drawn out into a refrigerated receiver by the vacuum system where it is held frozen.
- when the drying is over, the vacuum is broken using an inert gas like nitrogen. The material is allowed to cool before it is packaged.

14.3.1 Preliminary Freezing Prior to Drying

As freeze-drying involves sublimation of ice from the material being dried it is implied that the material should be in a frozen state before the start of drying. It is possible to achieve this by evaporative cooling by loading the material in the drying cabinet and drawing in a sufficiently high vacuum. In this case the water evaporates from the cut surfaces under low pressure absorbing the latent heat from the material eventually lowering the temperature to the freezing point. However, in certain cases it is considered desirable to pre-freeze the material before loading it into the dryer.

14.3.1.1 Size of Ice Crystals and Quality of the Freeze-Dried Product

The rate of freezing markedly influence the size of ice crystals formed, their location in the fish as well as the drying and rehydration characteristics. In quick freezing the ice crystals formed are extremely small in size and mostly inside the cells (intracellular). In a slow frozen product the size of the ice crystals becomes large and is more and more extracellular in location. This eventually can cause severe mechanical damage to the cell structure. As large crystals of ice sublime away, they leave behind pores of large diameter in the dry tissues offering reduced resistance to the escape of vapour from the ice surface as compared with a material that has been frozen rapidly. Such a product, on rehydration, will readily admit water to the centre of the piece almost instantaneously. However, with fibrous materials like fish and meat particularly, this water will be held very loosely only by capillary forces and can, therefore, be easily squeezed out. Furthermore, the structural changes accompanying slow freezing adversely affect the texture and appearance of the product. For many foodstuffs, therefore, it is considered desirable to adopt an intermediate rate of freezing as that obtainable in blast freezing which is likely to result in the formation of optimum size ice crystals.

Another factor influencing the quality of the product is the amount of water evaporating during the freezing step. This is of particular importance in the case of evaporative freezing where the food is kept in the drying chamber and subjected to very high vacuum. Since some of the water is evaporated under vacuum from the liquid stage, some solid migration and shrinkage of the material will occur. The solute transferred to the surface may hinder sublimation of ice. Under the vacuum used in the commercial freeze drying, the eutectic freezing point may not be achieved by evaporative freezing and hence part of the water may be continuously evaporated from the liquid state. Both these affect the quality of the product adversely.

14.4 FACTORS AFFACTING QUALITY OF FREEZE-DRIED PRODUCTS

The important factors affecting the quality and shelf life of freeze-dried products are the temperature attained by the product, its final moisture content and oxygen concentration.

14.4.1 Temperature of the Product and Ice During Drying

There are two types of heat input in freeze-drying viz., heat input through ice layer and heat input through dry layer. The former takes place in drying of liquids and the latter is associated with drying of solids.

In drying solids the highest temperature is attained by the surface of the dry layer in the product. This is controlled by the heat input to the product during drying. At the same time the temperature of the ice in the material in the receding zone of drying will be determined by the extend of vacuum maintained in the chamber.

The maximum temperature that may be tolerated largely depends on the nature of the food. It is, however, ensured that in any case severe scorching of the surface is avoided. Some of the factors to be considered in limiting the surface temperature are :

- destruction of pigments like astaxanthin/astacin in shrimp, myoglobin in meat etc.
- denaturation and cross linking of proteins.
- rendering out of fat. This will impart an ugly appearance besides causing imperfect dehydration due to formation of water-impermeable coating of fat on the internal surfaces of the food.
- non-enzymatic browning.

The maximum surface temperature found safe is 45-50°C for lean fish. In the case of crustaceans like shrimp it can go up to 60°C.

The limit of ice temperature is set almost exclusively by the temperature at which eutectic melting occurs. This depends on the concentration of solutes like sugars, salts etc. in the water. High sugar items are, therefore, processed at a low ice temperature, i.e., low chamber pressure.

14.4.2 Final Moisture Content

As a rule, freeze-dried foods are fully dried foods having residual moisture content as low as 1-2 per cent. As freeze-drying proceeds the sublimation front recedes and at a given time any undried portion will be located far removed from the surface. If the process is not complete, ice remaining in this location will melt and that portion of the food will remain similar to the original wet material. Spoilage can occur at this point (see Fig. 14.6) as in the original material irrespective of the extent of drying in total.

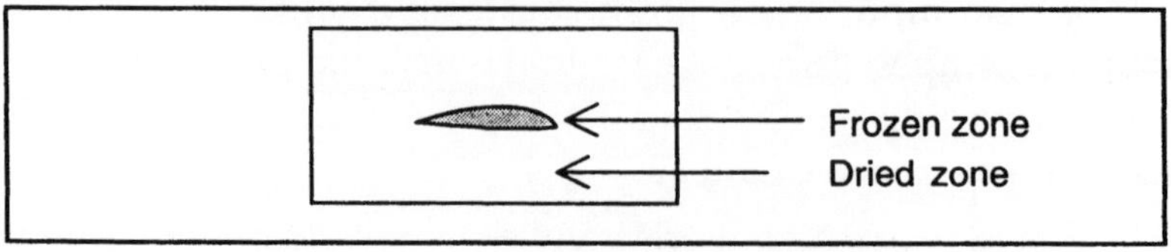

Fig. 14.6 : Location of undried portion

14.4.3 Oxygen Concentration

In normal freeze-drying process concentration of oxygen in the drying atmosphere is extremely low. However, at the completion of the process, if vacuum is broken by admitting air, oxygen may be absorbed by the material leading to difficulties for packaging it under inert gas. Oxygen can also adversely affect the colour and flavour of the product. The ideal solution to this problem is to break the vacuum in the chamber with nitrogen and to package under this gas without exposing the food to air. Removal of oxygen from the headspace of the package is an important pre-requisite for the stability of the freeze-dried foods. A generally accepted maximum level of oxygen concentration is about two per cent below as the atmospheric pressure.

14.5 ADVANTAGES AND DISADVANTAGES OF FREEZE-DRIED FISH

14.5.1 Advantages

Freeze-drying offers several advantages over the conventional air-dried fish :

- there is no product shrinkage as the process is a solid phase drying. Freezing effectively fixes the shape and, because ice is sublimed, fish retains the original shape and volume with crevices left from where ice sublimes.
- as ice directly sublimes there is no water movement and consequently no deposition of solids on the

surface. Therefore no case hardening as in air drying takes place in freeze-dried fish.

- as the process is carried out under very high vacuum and the sublimation front is always kept at very low temperature there will be no thermal degradation of proteins or deteriorative changes in colour or flavour.
- freeze-dried foods rehydrate rapidly by simple immersion in water for few seconds to few minutes. Water quickly enters the cervices left by sublimation of ice. Rehydration up to 95 per cent is possible.
- freeze-dried foods in ideal packaging are very stable at ambient temperatures. No cold storage is needed for freeze dried foods.
- no cold chain distribution system is needed, which obviate the need for refrigerated transport.
- freeze-dried foods have a 'convenience' factor with regard to the easiness with which it can be made ready for the table.

14.5.2 Disadvantages

In spite of the several advantages associated with freeze-dried foods some disadvantages also are apparent with respect to the economy of the process and some related to storage and distribution of the products. Important among them are :

- needs sophisticated and expensive equipment.
- the energy requirement is high.
- the product on rehydration holds water by capillary forces only. The absorbed water can be easily squeezed out.
- drying should proceed till the moisture level is as low as 1-2 per cent.
- product is porous and fragile. Sharp contours can rupture film packages necessitating sturdy packaging.
- product, often, is prone to oxidation. Therefore, it should be packaged under vacuum/inert gas.

14.6 PACKAGING

Packaging influences the shelf life of freeze-dried fish to a great extent. The important factors which determine the type of packaging necessary for freeze-dried foods are

- uptake of oxygen.
- uptake of moisture.
- flavour contamination.
- contamination by microorganisms and insects.
- exposure to visible and ultraviolet light.

Fish and shellfish products are very sensitive and hence require maximum protection against oxygen, water vapour, light, mechanical damage as well as protection against contamination. The important requirements for packaging such products are that the packages should not virtually allow any diffusion of water vapour and oxygen.

Prevention of migration of atmospheric moisture and oxygen, exposure to light and, microbial contamination can readily be solved if hermetically sealed cans are used as a container. In many cases this type of packing is used for commercial packaging of freeze-dried products, particularly fish.

Flexible pouches made of aluminium foil laminated with plastic materials offer another important group of packaging materials suitable for freeze-dried foods. Flexible pouches even offer certain additional advantages over rigid packaging. Freeze-dried products are fragile and susceptible to breakage and fragmentation. Some breakage and "dusting" invariably take place in products like shrimp, lobster etc. during storage and transport. Some improvement in this aspect can be achieved by packing such foods in flexible packaging under nitrogen with internal pressure few centimetres below the atmospheric. The product is given a 'snug' fit inside the pouch and there is little breakage due to impact of individual pieces of the product. Impact damage due to shock to the pouch as a whole is minimised by placing the product inside a rigid container.

The packaging detailed above are, perhaps, the best for freeze-dried foods. In addition to acting as good barrier against oxygen, moisture etc. they also protect the food from light and microorganisms.

14.7 CHANGES TAKING PLACE IN FREEZE-DRIED FISH DURING STORAGE

Freeze-dried foods, fish in particular, undergo deteriorative changes during storage. The main deteriorative changes taking place are :

- lipid oxidation.
- oxidation of other components.
- non-enzymatic browning.
- changes in the proteins and other polymeric constituents.

The deteriorative changes taking place in the products depends on

- the nature of the food and its treatment prior to storage.
- time and temperature of storage.
- oxygen concentration.
- moisture content.
- light intensity.

14.7.1 Lipid Oxidation

This is one of the most critical problems limiting the shelf life of freeze-dried products, particularly fish products. The hydroperoxides formed as a result of oxidation may further breakdown to products which can impart undesirable flavours and, even toxicity. Peroxides may react with fat soluble vitamins, pigments and flavour compounds. Some of the oxidation products can react with amino acids thus reducing the protein quality. Exclusion of oxygen or at least minimising the oxygen pressure within the package, exclusion of light and, incorporation of antioxidants are effective preservative measures against lipid oxidation.

14.7.2 Oxidation of Other Constituents

Oxidation is not limited to lipids. Oxidation and resultant discolouration is of great significance in products like shrimp which contain only very little lipids. Astacene pigment in shrimp will undergo oxidative deterioration yielding deep yellow discolouration products.

Interaction with molecular oxygen can damage proteins. In the case of haemoglobin and myoglobin brown discolouration takes place because of the oxidation of the iron-containing haem group attached to the protein. This is of great significance in products like meat.

14.7.3 Non-enzymatic Browning

Non-enzymatic browning is a common problem in freeze-dried fish and meat which is caused by the formation of di-, and polycarbonyl compounds, which polymerise to yield brown compounds. Reduction of moisture to very low levels is the most effective way to prevent this reaction in freeze-dried products.

14.7.4 Changes in the Protein and Other Polymeric Constituents

If the temperature of storage is high and the dried product has a moisture content above two percent, the water holding capacity which is already adversely affected during drying will become more aggravated. The changes are associated with cross-linking of polymeric structural components like protein in fish and meat. It is more or less understood that cross-linking of myofibrillar protein is the most important mechanism involved. Cross-linking of proteins will lead to increased toughness of the product.

14.8 REHYDRATION

Freeze-dried foods rehydrate quickly. Rehydration can be evaluated in terms of total water absorbed by the food or in terms of water holding capacity which is a measure of water retained by the food after application of pressure. The degree of rehydration will depend on several factors like time, temperature and nature of the rehydrating medium.

The moisture content of rehydrated freeze-dried foods as also their water holding capacity are always somewhat lower than those of the raw materials from which they are processed. Temperature is known to play an important role in rehydration. Rehydration is more at lower temperature than at higher temperature. The lower water uptake of freeze-dried fish at higher temperature may be due to the denaturing effect of higher temperature on protein or the movement of fat in the hot rehydrating fluid with blocking of the interstitial channels formed during freezing.

Suggested Reading

Brennam, J.G., Butters, J.R., Cowell, N.D. and Lilley, A.E.V. (1990) *Food Engineering Operations* 3rd edn. Elsevier Applied Science, London and New York.

Charm, S.E. (1978) *The Fundamentals of Food Engineering.* AVI Publishing Co., Westport, Connecticut, USA.

Cotson, S. and Smith, D.W (eds) (1962) *Freeze-drying of Foodstuffs.* Columbine Press, Manchester and London.

Dalgleish, J.Mc.N. (1990) *Freeze-drying for the Food Industries.* Elsevier Applied Science, London.

Heid, B.S. and Joslyn, M.A. (1981) *Fundamentals of Food Processing Operations.* The AVI Publishing Co., Westport, Connecticut, USA.

H.M.S.O (1960) *Accelerated freeze-drying methods of fish preservation.* Her Majesty's Stationery Office, London.

15
IRRADIATION

SEVERAL methods are employed for preservation and extension of shelf life of fish ranging from primitive drying/ smoking to freezing and freeze-drying. Another important step forward is the development of technology for transportation of live fish. A notable and conceptual difference from all these methods is utilisation of ionising radiation for food preservation. Preservation of foods using ionising radiation is called irradiation. Preservation of food by irradiation is one of the truly peaceful uses of atomic energy. Irradiation of foods has been found useful and effective to

- inhibit sprouting or reducing weight losses in vegetables such as potato, onion etc. during storage
- delay the ripening of fruits
- kill insect pests in fruits, grains or spices
- reduce or eliminate food spoiling microorganisms in meat and seafood products

15.1 IONISING RADIATION AND PRESERVATION OF FOODS

High-energy radiation is capable of inactivating micro- and other organisms. Wavelength of high-energy radiation is much shorter than that of visible light spectrum. Because of shorter wavelength and higher energy, these radiation have a higher penetrating power. Ionising or high-energy radiation induces chemical changes in the individual compounds of the food. Such changes often involve decomposition reactions giving rise to new chemical species.

Ions and excited molecules are the first species of new chemicals formed when ionising radiation is absorbed by the material. The free radicals formed by the dissociation of excited molecules and by ion reaction are largely responsible for the chemical changes observed in the food.

In food, the major effect is splitting water molecules to yield free radicals and hydrogen peroxide. Free radicals are highly reactive from which the useful effects of irradiation originate. They do not persist in aqueous systems, but rapidly recombine. Irradiation brings about modifications in the DNA of the bacteria such that it results in the death of the bacteria or their inability to reproduce. DNA molecules are far more sensitive to radiation than any other molecule in the food undergoing irradiation. As a result the living organisms such as bacteria, yeast, moulds etc. are killed long before any undue change takes place in the food. Thus food spoiling bacteria and pathogens in food can be destroyed by irradiation without causing any other serious change in the food. Parasites as well as insects also can be killed at very low doses of irradiation. Provided the process is adequately controlled, no possible health hazard can be foreseen from consumption of irradiated food. Irradiation can, therefore, be used to improve food safety, minimise losses and extend shelf life.

15.2 SOURCES OF IONISING RADIATION

There are two main sources of ionising radiation commonly employed in food preservation. These are gamma rays and electron accelerators.

15.2.1 Gamma Rays

Gamma rays are electromagnetic radiation of a very short wavelength produced by the spontaneous disintegration of the atomic nucleus of certain radioactive nuclides, e.g., cobalt-60, caesium-137. Both are man-made gamma rays emitting materials. Cobalt-60 is an isotope of cobalt produced by irradiating cobalt in a nuclear reactor, and caesium-137 is present as a fission product in the fuel elements used in nuclear reactors. Cobalt-60 is the most commonly used source of gamma rays in food irradiation. Gamma rays have very high penetrating power and can, therefore, be used to irradiate bulk items like products in their final shipping container.

15.2.2 Accelerated Electron Beam Irradiation

Electron accelerators produce high-energy electron beams and accelerate them to very high speeds producing very high levels of energy in fractions of a second. This process is a non-nuclear method and, unlike in gamma rays, the source can be switched off when not in use and hence is easy to control. Electron beam irradiation is also quicker than gamma irradiation. However, the degree of radiation penetration is poor and hence its application is limited to irradiation of thin layers of foods.

A third source of ionising radiation is the X-ray. X-rays are produced by conversion from electron beams, but the conversion is not efficient, is uneconomic and hence not in use now for irradiation purposes.

15.3 FOOD IRRADIATION PROCESS

Irradiation is carried out using a high-energy source in a chamber provided with thick concrete walls. The energy source is the cobalt unit having a specific area activity more than 100 curies per gram. This is assembled in the form of tiny cylinders to form rods 450 mm long. These rods are fully screened by hermetically welded double walls of stainless steel. A rod of cobalt-60 usually contains 10,000 curies. A vessel filled with water serve as a shield against irradiation when not in use. The food, packaged and frozen or chilled, is loaded onto a conveyor belt that passes into the irradiation unit. There should be no contact of the food with the energy source. The speed of the conveyor is so adjusted that the food remains exposed to the source for a predetermined period. Effect of irradiation depends on the dose received by the food which, in turn, is decided by the residence time in the irradiation chamber. Since the power of any source is constant, this is the only variable to decide the dose of radiation. The food leaving the unit can be handled in the normal way for storage and distribution. A typical layout of a gamma irradiation unit is illustrated in Fig. 15.1.

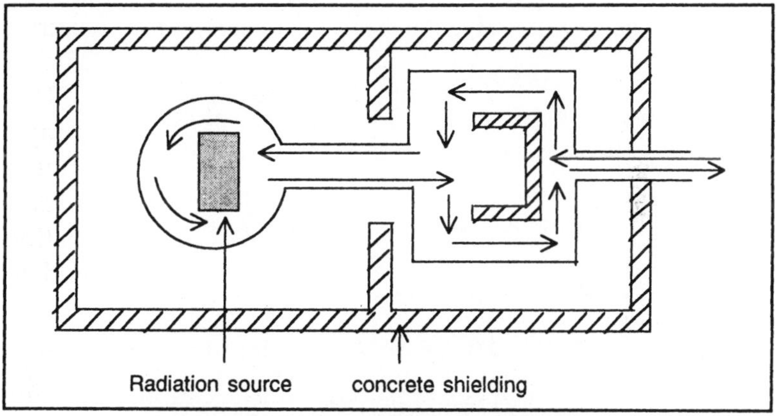

Fig. 15.1 : Set up for food Irradiation

15.4 MEASUREMENT OF RADIATION ENERGY

The unit earlier used to measure the dose of radiation was 'rad'. A radiation dose of one rad involves the liberation of 100 erg of energy into each gram of matter through which the radiation passes.

1 erg = 10^{-7} joules

The SI unit now used to measure radiation dose is the gray (Gy).

The energy of all types of radiation is measured in electron volts, eV, or more normally, Million eV (10^6 eV), MeV.

Electron volt is a unit of energy equivalent to the amount of energy gained by an electron when accelerated by a potential of one volt.

1 eV $= 1.6 \times 10^{12}$ erg

1 Joule $= 1 \times 10^{7}$ erg.

A dose of 1 Gy is the absorption of 1 J of energy per kg of food

1 Gy = 1 J/kg

1 kGy = 1000 Gy

15.4.1 Radiation Dose

Different doses of radiation are applied to foods to achieve different purposes. Extension of shelf life of fish and fish products are generally achieved by reduction in microbial loads. This is achieved when the radiation dose is between 0.5 and 5.0 kGy. Elimination of non-sporing pathogens may need a radiation dose of 3 to 10 kGy and bacterial sterilisation, upto 50 kGy.

15.5 IRRADIATION OF FISH AND SHELLFISH

There are three main areas where irradiation is helpful in fish preservation. These are : disinfestation, extention of storage life and destruction of pathogens.

15.5.1 Disinfestation

Disinfestation becomes useful in storing dried fish. Insects infesting dry fish can be destroyed by a radiation dose upto 1 kGy. However, it is essential that the fish stored have been appropriately packaged to prevent re-infestation. Parasites also can be destroyed by very low doses of radiation.

15.5.2 Extention of storage life

Employing doses upto 2 kGy, the numbers of food spoiling organisms can be significantly reduced, thus resulting in extension of shelf life.

15.5.3 Destruction of pathogens

Pathogenic bacteria like Salmonella and Listeria can be destroyed with radiation doses between 2 and 5 kGy. These doses are sufficient to destroy only viable cells; higher doses will be necessary to eliminate bacterial spores. High doses will be required in the case of frozen foods also. The effect of irradiation is dependent on the availability of free water, and much of the water in frozen fish is bound as ice.

15.6 RADIATION PASTEURISATION

Two broad microbiological objectives can be distinguished in preservation of food by irradiation. One is total destruction of all the pathogenic and spoilage microorganisms. This can be called 'radiation sterilisation'. The second is the elimination of the pathogens and also reducing the viable organisms, in general, with consequent improvement in the keeping quality of the food.

Irradiation process is accompanied by an increase in temperature, though it is not very significant. Even a sterilisation dose may increase the temperature only marginally, say by 10°C. But the overall effect on quality is often extreme. Therefore, the interest is for use of lower doses. At the lower doses, irradiation only pasteurises the food. Therefore it is necessary to hold such foods at lower temperatures to prevent the remaining microorganisms from multiplying and spoiling them. This is often referred to as 'radiation pasteurisation'

15.7 EFFECT OF IRRADIATION ON FISH AND FISH PRODUCTS

15.7.1 Disinfestation

As stated in Section 15.5.1, low dose irradiation is a powerful means of controlling insect infestation in dried fish. The effect will depend on the moisture content of the product and the dose employed. When the moisture content is below 20 per cent, an irradiation dose of 0.15 kGy can prevent the larvae of insects from developing into adults. At 0.25 kGy, the larvae will survive, but stops feeding. For products with moisture content between 20 and 40 per cent, which is the general moisture range in dried fish, a minimum dose of 0.5 kGy is recommended to control damage by flies and insects.

15.7.2 Effect on Microorganisms

Irradiation produces chemical changes that, in turn, produce biological changes in the foods. The major benefit of irradiation is its effect on microorganisms. The effect on microorganisms will depend on the dose of irradiation employed and, with respect to doses employed, the processes are termed radurisation, radicidation and radappertisation.

Radurisation is the process of irradiation where low doses, usually below 1 kGy, is employed. These doses will extend the shelf life by (a) inhibiting sprouting in vegetables like potato and onion, (b) delaying ripening of fruits, (c) eliminating insect pests in grains, spices etc.

Radicidation refers to irradiation employing doses between 1 kGy and 10 kGy. Irradiation in this range results in decrease in the number of food spoiling microorganisms like yeast, moulds and non-spore-forming bacteria and thus extends the shelf life of foods or reduce the risk of food poisoning. Useful reduction in counts can be achieved in lower doses of, say 2 kGy.

Radappertisation involves use of high doses above 10 kGy and upto 50 kGy. These extremely high doses can effectively sterilise the foods. However, the high dosage levels will induce undesirable changes in flavour and texture.

15.7.2.1 Mode of Inactivation of Microorganisms by Irradiation

Irradiation inactivates microorganisms predominantly because of the charged particles generated by it. Although these charged particles are short-lived, they can cause cleavage of the deoxyribonucleic acid (DNA) in the microorganism. The DNA which contains the genetic information necessary to create progeny must remain intact if the microorganism is to be capable of growth and multiplication. The cleavage caused in the DNA due to irradiation inactivates the bacteria.

15.7.2.2 *Extent of Inactivation of Bacteria*

The dose of irradiation required to achieve the desired microbiological quality of the food depends not only on the type of organisms involved, but also on the number of organisms likely to be present in the food before irradiation as also the number of viable organisms that can be tolerated in the irradiated product. Larger their initial number and smaller their final number to be attained, the higher will be the radiation dose required. The microorganisms considered to be of public health significance are the same whether in irradiation process or processes other than irradiation, for example, *Clostridium botulinum, Bacillus cerus, Salmonella, Staphylococcus aureus* etc.

With radiation doses up to 10 kGy a high rate of destruction of bacteria takes place. Most of the pathogens like *Shigella*, enteropathogenic *E.coli, Streptococcus faecalis, Staphylococcus aureus, Salmonella typhi, S. paratyphi* etc. are destroyed in this range. However, enzymes are not inactivated by the radiation treatment and enzyme-linked spoilage may predominate in irradiated foods. Therefore, the terminal spoilage in irradiated seafood is often characterised by a sweetish, slightly spoiled or fermented flavour and odour, which are markedly different from the commonly experienced putrid smell in non-irradiated fish.

Though irradiation in the above dose will kill bacteria, still higher doses in excess of 25 kGy will be needed to eliminate the bacterial spores. Therefore, spores of *Clostridium botulinum,* a potential lethal toxin-producing bacteria, will survive such treatments. *Clostridium botulinum* is also the most radiation-resistant, spore-forming spoilage organism known Therefore fish, which are potential carriers of spores of C. *botulinum*, should be stored below 3.3°C, the temperature at which it can produce toxin, after irradiation. This is of greater significance because the low dose irradiation sufficient to kill the spoilage bacteria will reduce the competition for the growth of C. *botulinum* and increase its multiplication after irradiation and storage.

Foodstuffs are generally irradiated after its final packaging. An unpackaged irradiated food will be only as good as an unirradiated food because of the possibilities of recontamination after leaving the irradiator and during further handling.

15.7.3 Effect on Nutrients

Irradiation does not bring about very significant changes in the nutritive value. Compared to some other processes involving heat treatment, or drying the changes in the nutrients in irradiated foods are less. Proteins and aminc acids are little changed at doses upto 10 kGy and are not affected to the extent that they lose the nutritional qualities.

Among the micronutrients present in fish, minerals are unaffected by irradiation, but there occur some losses in vitamins and polyunsaturated fatty acids. Some vitamins like vitamin A, E, K and thiamin are sensitive to irradiation. Other vitamins like riboflavin, niacin and vitamin D are relatively insensitive. Vitamin C level decreases during storage, but is of no significance in fish.

15.7.4 Organoleptic Changes

Irradiation causes only marginal changes in flavour, texture and odour of fish compared to those brought about by heat processing. Products maintain the fresh quality and are indistinguishable from untreated products. However, with fatty fish development of rancid flavour is a problem.

This is of serious concern in fish containing highly unsaturated fats. The free radicals produced as a result of irradiation can initiate autoxidation chain that will lead to rancidity. Oxidation of unsaturated fats is likely to cause a reduction in their nutritional advantages. In such cases it is desirable to apply a dose which, while achieving desired antibacterial activity, can maintain an acceptable organoleptic quality. Integrating packaging under vacuum with irradiation also can suppress bacterial spoilage and oxidative rancidity.

The radiation-induced flavour changes, often described as 'metallic', 'burnt feather-like', 'rubbery' etc., are associated with irradiated fish even when the dosage is low. With sterilisation doses, strong bitter flavours develop in fish. Some other organoleptic changes associated with sterilising doses are brown and other types of discolouration, and toughening of texture. All these will become accelerated during subsequent storage at temperatures in the range 0-15°C.

Some fat-based pigments get bleached during irradiation. Generally irradiated fish do not show any increase in TMA or TVB during storage, perhaps because the bacteria producing these substances are destroyed.

15.7.5 Preformed Bacterial Toxins

Food products may contain bacterial toxins if they have been contaminated with toxigenic bacteria and held at conditions favouring development of toxins. Effect of irradiation on these toxins differs from that of heat. The toxin of *Clostridium botulinum* is readily destroyed by heat, but not by radiation at the dose levels used for food preservation. The staphylococcal enterotoxin cannot be destroyed with certainty even by extreme treatment with either heat or radiation. Therefore, the food capable of supporting the development of toxin should be irradiated only if it has been produced and stored under good sanitary conditions.

15.7.6 Packaging of Irradiated Food

Packaging materials used as containers for irradiated foods should be subjected to careful scrutiny to ensure their suitability and safety for use. In contact with food, certain extractives from the package may contaminate the food. Therefore, before a packaging material is recommended, consideration must be given to the possibility of migration of extractives from it into the food. Extractive studies on such materials using food-simulating solvents should be conducted before and after irradiation at the appropriate dosages.

Flexible packages have special advantages and many plastic materials can be used in conjunction with irradiated foods. Ionising radiation may also be used to improve the properties of certain polymeric packaging materials. It may also be used to inactivate any microorganism, which the packaging material might have been contaminated with prior to bringing into contact with the food.

The packaging material used should also be such that it does not get itself irradiated and induce radioactivity in the food contained.

15.7.7 Supplementary Treatments for Some Irradiated Foods

Sometimes it may become necessary to resort to some treatments in addition to irradiation to ensure the keeping quality of the food. For example, destruction of enzymes requires an irradiation dose that will render the food unacceptable. Therefore, some pre- or post-irradiation treatment for the destruction of the enzymes becomes necessary. Heating may be used in conjunction with irradiation to inactivate the enzymes.

15.8 DETECTION OF IRRADIATION IN FOODS

Irradiation produces little or no change in the appearance or taste of the food if it is within the limits. Therefore, it is difficult to distinguish an irradiated food from a non-irradiated one. The problem of detection becomes still more complicated if an irradiated food product is used as an ingredient in some combination. No generally accepted method is known at present which can be used to indicate whether or not the food is irradiated and, if irradiated, how much radiation has been applied.

The free radicals produced as a result of irradiation do not persist in aqueous systems as they rapidly recombine. However, they persist in hard and bony materials such as the shell of shellfish. This persistence is being researched for detection of irradiation in shellfish. The technique is called Electron Spin Resonance Spectroscopy. Some of the other physical methods considered applicable in the case of fish are polarographic and polarometric changes and, changes in the rH potential.

Chemical methods which offer possibility in detection of irradiation of foods are : (i) the changes in organoleptic quality detectable by gas chromatography or the appearance of specific thermoluminescence and (ii) colourimetric detection of short-chain aldehydes.

15.9 TOXICOLOGICAL ASPECTS OF FOOD IRRADIATION

Before permitting irradiation in food preservation, it is extremely important to decide whether consumption of irradiated foods could lead to short or long-term toxic effects. The evaluation of toxicological safety can be done by

- evaluating the toxicity of individual radiolytic products detected in foods.
- considering effects in experimental animals where whole diet is irradiated, and also where individual irradiated food items have been fed to animals.

The most appropriate way of determining the overall toxicological consequences of consuming irradiated foods is the studies on mutagenic hazards. These studies have shown that no toxicological hazard to human health would arise from consumption of food irradiated up to an average dose of 10 kGy. There is no evidence of toxicological hazard associated with higher doses; however, the recommended dose is upto 10 kGy.

15.10 LEGISLATION

Several countries have permitted irradiation in foods with respect to specified purposes. Only very few countries have permitted it in fish and have specified the dosages and purpose of irradiation. There is no uniform attitude towards irradiation.

The F.A.O./W.H.O. have drawn up a Codex General Standard for irradiated foods. Guidelines for the choice of source of radiation, facilities needed, control measures, labeling etc. have been provided. A maximum dose of 10 kGy and at energies not exceeding 10 MeV for accelerated electrons or 5 MeV for gamma radiation has been recommended for irradiating foods. This will not lead to any significant change in the natural radioactivity or prejudice the safety and wholesomeness of the food.

Labelling is considered desirable to inform the consumers that the food has been irradiated and to provide instructions for its handling and storage.

Suggested Reading

Coleby, B. and Shewan, J.M. (1965) The Radiation Preservation of Fish. In : *Fish as Food.* Vol. 4, Part 2, Borgstrom, G. (ed.), Academic Press New York.

Desrosier, N.W. and Rosenstock, H.M. (1960) *Radiation Technology in Food,* The AVI Publishing Company, Westport, Connecticut, U.S.A.

FAO (1965) *The Technical Basis for Legislation on Irradiated Foods.* FAO, Rome.

FAO/IAEA (1970) *Preservation of Fish by Irradiation.* International Atomic Energy Agency, Vienna.

FAO/WHO (1983) *Codex General Standards for Irradiated Foods.* FAO, Rome.

FAO/IAEA (1989) *Radiation Preservation of Fish and Fishery Products.* Technical Report Series No. 303 FAO, Rome.

Kreuzer, R. (ed.) (1969) *Freezing and Irradiation of Fish.* Fishing News Books Ltd., Byfleet, Surrey, UK.

Thorne, S. (1991) *Food Irradiation.* Elsevier Applied Science, London and New York.

Urbain, W.M. (1978) In : *Advances in Food Research,* Vol. 24, Chichester, C.O. (ed.) Academic Press, New York.

16
TRANSPORTATION OF LIVE FISH

SUPPLYING fish live to the consumer is probably the best way to ensure that the fish he gets is fresh. However, until recently commercial transportation of live fish was almost confined to ornamental aquarium fishes because of their immense commercial potential in export as well as domestic markets. With the development of aquaculture of fish and shellfish, live fish transport assumed greater significance primarily in transporting live seeds from hatcheries to culture farms. This was closely followed by transport of live food fishes.

There have always been ever increasing and ever changing demands for diverse types of processed fish and fish based products from consumers all over the world. With changing life styles of people, the demand from many of the affluent consumer markets gradually started shifting towards live fish and shellfish and this is showing an increasing trend. Even though live transport of fish;, particularly of farmed freshwater fish, was prevalent in Europe, transport of several species of fish and shellfish - marine, brackish water and freshwater - to distant markets is relatively of recent origin. Affluence of the consumers and the scientific understanding and technological developments in handling and packaging techniques have provided added impetus to the prospects of commercial transportation of live fish and shellfish for human consumption.

Three common methods employed in transportation of live fish are :

- **waterless method** - usually practiced in the case of molluscan shellfish like clam, mussels, oysters and, crustaceans like some species of shrimp, lobster and crab. The specimens are held in a cold moist atmosphere for the duration of transport
- **plastic bag method** - usually adopted for live transport of ornamental aquarium fishes and fry/ larvae of freshwater fish.
- **the tank method** - in which the fish is kept in water in tanks of various types and often kept open to the atmosphere.

The details will be discussed later in this chapter.

16.1 LIVE FISH TRANSPORT - SOME GENERAL PROBLEMS

Handling live fish without appreciable mortality until it reaches the consumer poses several problems, some of the important among them being :

- low solubility of oxygen in water together with its poor capacity to dissipate the end products of metabolism
- need of extreme care in handling. Live fish is very sensitive to handling. In some delicate species, the fish will lose the essential protection from osmotic stress if the mucus is removed from even a fraction of the area of the skin by abrasion or some other means
- capture and subsequent handling cause the fish to be excited and stimulated to the extent they readily accumulate dangerous levels of lactic acid in their blood
- excessive changes in the temperature from catching to handling and live carriage is deleterious to the fish in many ways

16.2 FACTORS AFFECTING SUCCESSFUL TRANSPORAT OF LIVE FISH

The important factors affecting the successful transport of fish live are :

- feed.
- respiration and the resultant metabolites.
- water temperature.
- use of anaesthetics.
- capturing and handling techniques and their effect on the fish quality.
- conditioning the fish to the packaging and transporting atmospheres.
- efficiency of the packaging and transporting systems.

16.2.1 Feed

Normally, the fish have to be maintained live only for short periods for live transport. In live carriage the fish will be crowded in the limited quantity of water available and, if fed under this condition, the water will become rapidly polluted due to the accumulation of metabolic wastes, making high oxygen demands. Normally the fish are not fed during transport. Feeding is often stopped 24 hours before packaging and transport so that the gut is empty before transport begins, and the release of metabolites to the water is minimised. Therefore, in practice, feeding is not of any serious concern in transport of live fish.

16.2.2 Respiration

16.2.2.1 Oxygen Needs

Though the feeding can be dispensed with even before packaging for transport, the fish must respire in order that they remain live. Therefore, it has to be ensured that an adequate supply of oxygen is provided and that this oxygen is in solution. One of the serious limiting factors regarding the number of fish to be carried in a container is the poor solubility of oxygen in water. Aeration can improve the supply of oxygen, but it becomes necessary to remove the carbondioxide produced in respiration as also other metabolic wastes such as ammonia.

Even though the fish may be in a sort of sedation while in the package for transit, it should at least perform compensatory movements throughout the transit period. Moreover, the fish is subjected to severe stress while catching. The struggle during harvesting and handling of fish is often violent enough that their oxygen consumption will increase and will require more of it than is available in the muscle. Under these circumstances, lactic acid builds up in the muscle due to anaerobic breakdown of carbon compounds and a condition of 'oxygen deficit' is created. It may take several hours to reduce the 'oxygen deficit' to zero even after the animals calm down. Therefore during struggle and a long time thereafter the oxygen usage is very high. If the fish has severely struggled during capture the blood pH may fall to such levels enough to cause death.

Replacement of air with oxygen may increase the survival time of fish. However, the packing density, which is directly related to the consumption of oxygen also, is a factor. Crowding of too many fish in limited volume of water also will reduce the oxygen availability. Packing density should be such that there will be minimum mortality during normal transit period of 3-4 days. Lower water temperature, use of anaesthetics and starvation will reduce the metabolic activity of fish and, thus, the oxygen demand will become lower. The rate of consumption of oxygen is reduced with increase in age and body weight, large fish consuming less oxygen per unit body weight than do smaller ones.

16.2.2.2 *Release of Carbondioxide*

The process of respiration releases considerable quantities of carbondioxide. For every millilitre of oxygen consumed by the fish, it will produce approximately 0.9 ml of carbondioxide. Carbondioxide produced may enter into equilibrium systems of carbonates, bicarbonates and also as free carbondioxide dissolved in water. Increase in the level of free carbon dioxide depresses the ability of the active fish to take up dissolved oxygen from water and thus becomes toxic to fish. This is further aggravated by an increase in temperature. There is always a limit upto which the fish can tolerate accumulation of free carbondioxide. It is up to 25 ppm for sensitive species and up to 50 ppm for insensitive species under adequate aeration. When the levels exceed these limits any increase in the amount of oxygen, provided by aeration or otherwise, will have no beneficial effect. Increased acidity of the water through accumulation of carbondioxide results in higher metabolic rates and also causes mortality in fish. Hard water is preferred for transportation of fish because it has greater capacity to buffer the free carbondioxide produced by the animals. Carbon dioxide concentrations above 20-30 ppm in the air supplied to transport tanks can cause severe oxygen stress. If pure oxygen is supplied, higher carbondioxide concentrations can be tolerated. Buffers such as tris may be used to maintain pH at the required levels. In addition to this, free carbon dioxide will attack unprotected metals. Zinc or copper from galvanised piping or copper tubing in the transport tanks may get dissolved and become toxic to fish. Therefore copper and zinc should be avoided in the construction of water tanks and equipment in contact with water.

16.2.2.3 Accumulation of Ammonia

About half of the nitrogenous excretion of the fish consists of ammonia. Ammonia is also produced by bacterial breakdown of organic compounds. Free ammonia or ammonium hydroxide is highly toxic to fish. The rate of excretion of ammonia is related to the rate of metabolism. The rate of excretion is also related to the size of the fish. Large fish of a given species produce less of excretion products than do smaller ones of the same species. Whereas, in a typical case, a fish weighing 20 g excreted nitrogen at the rate of 500 mg/kg/day, a fish weighing 100 g excreted nitrogen at the rate of 120 mg/kg/day only. Large fish are also somewhat more resistant to the toxicity of ammonia than are smaller fish. At ammonia concentrations as low as 0.5-1.0 ppm only non-lethal toxic effects take place. The oxygen demand of the water, other than that required for direct respiration of fish, will be higher if the fish are kept longer in water because of the accumulation of organic metabolic wastes.

Discharge of metabolic waste may be controlled by lowering the metabolic rate of fish and using suitable substances to remove them. Reduction in metabolic rates can be achieved by lowering the water temperature, addition of anaesthetics to water, and through conditioning of the fish.

16.2.3 Water Temperature

For live transport, fish has to be packed in water at cold temperatures. For cold water species this is not a problem; but problems arise in live transport of tropical species, as they cannot tolerate low temperatures. Increase in temperature causes increased metabolic activity of fish and bacteria, which increases oxygen demand and production of ammonia. Warmer water can hold only less oxygen than cold water. The affinity of the blood for oxygen decreases when the temperature increases. Carbon dioxide and ammonia are more damaging at higher temperatures.

Sudden changes in temperature will cause shock to the fish, a shock resulting from the failure of the nervous system and another resulting from the failure of a physiological function such as respiration. The fish suffering from shock will exhibit symptoms like surfacing and gulping for air, loss of equilibrium, loss of orientation etc.

Temperature shock can occur when there is a rapid change of 5°C in the temperature. Therefore for warm water species temperature change exceeding 5°C should be done over several hours. Fish acclimate more rapidly when changed to a higher temperature than when moved to a lower temperature.

16.2.4 Overexertion of Fish

During catching, as also during subsequent handling, the fish may struggle very much. Struggling may be to avoid capture, or in the live box during handling, packaging, and transport. However, struggling will involve severe muscular activity and even death may occur as a consequence. During struggling much of the stored glycogen gets converted to lactic acid causing severe acid-base disturbances, and this may be the cause of death. At high acid concentration the oxygen carrying capacity of the blood will become seriously affected. The ability of the heart to pump blood at appropriate levels also will be reduced as a consequence of struggling. This may also reduce the swimming capacity of the fish. Thus, a struggled fish going into a live carriage will be at a disadvantage aside from any death occurring directly as a result of overexertion.

16.2.5 Bacterial Population

As organic matter builds up in the transport tank, the bacteria present in the fish and the tank may multiply rapidly. Some bacteria may use ammonia as food and thus reduce the ammonia concentration, but this is not significant. Aeration and periodic water changes are the possible control measures. Antibiotics have limited value in controlling the bacterial population in transport tanks as they appear to kill only susceptible species while allowing others to multiply rapidly.

16.2.6 Use of Anaesthetics

Anaesthetics are used to reduce the metabolic rate of fish. Chloral hydrate, tertiary methyl alcohol and methyl parafynol are some of the most effective anaesthetics to induce quiescence in fish during transport. Sodium amytal, often used as an anaesthetic will be affected by the calcium content of water. Tricane methane sulphonate, though tends to breakdown, has been successfully used in many cases of live transport of freshwater fish.

The anaesthetic should only produce a stage of sedation in fish and not beyond that. Therefore the doses should be carefully chosen. Smaller fish are more resistant to anaesthetics than are larger specimens. Rate of induction of anaesthesia is accelerated at higher temperatures but there is little temperature effect on the level used.

16.2.7 Use of Antiseptics and Antibiotics

In some cases some prophylactic or quarantine measures may become necessary because of the possibility of induction of infectious diseases, parasites etc. in fish consignments. Antiseptics, antibiotics and germicidal chemicals may be used for this purpose. A short duration chemical bath prior to transportation will help in preventing further transmission of infections in fish consignments. Copper suphate, methylene blue, potassium permanganate, sodium chloride, acriflavin and chloromycetin are few of the commonly used chemicals.

16.2.8 Conditioning

The environment in which live fish is transported is different from the natural environment from which it is caught. Therefore, immediate transfer of the fish to the transport medium will cause severe stress increasing the metabolic rate and fouling during transit. Therefore, the fish is subjected to conditioning so that they get adapted to the new environment. The fish are transferred to tanks filled with well-aerated water to ensure that there is no sudden change of pH and no sudden temperature shock. Conditions in the tank during transport are maintained as close as possible to the optimum for a particular species.

Several chemical substances are used for absorption and removal of metabolic wastes. Clinoptiolite or activated charcoal, through selective ion exchange, also removes ammonia from freshwater live fish transport medium. For seawater transport nitrifying bacteria can be employed for oxidising ammonia and its removal.

16.2.9 Use of Antifoaming Agents

Foam production has been experienced in the live transport of certain species of fish. Thick foam develops from the mucous and organic matter affected by aeration. The foam will cover the water surface reducing the availability of oxygen and resulting in accumulation of carbon dioxide and the consequent acidification of water. Foaming can be eliminated by the use of antifoaming agents.

From the foregoing it becomes apparent that the important factors to be considered in transportation of live fish are

- condition of the fish prior to packaging.
- conditioning the fish for a period of time before packaging or transport to acclimatise with the environment in package and transport.
- maintenance of low storage temperatures.
- maintenance of natural environmental conditions.
- careful handling to avoid stress and injury to fish.
- appropriate stocking density and water temperature.
- duration of holding in the package.
- replenishment of air with oxygen.
- maintenance of high humidity in the packages for molluscs and crustaceans.
- reduction in accumulation of toxic metabolites and controlling the acidity of water using suitable buffers
- use of antibiotics/antiseptics/antifoaming agents wherever necessary.

16.3 METHODS OF TRANSPORT OF LIVE FISH

Depending upon the type of fish/shellfish transported, three different methods are in common use. These are the water-less method, plastic bag method and the tank method.

16.3.1 Waterless Method

Many molluscan shellfish like clams, mussels, oysters etc. and crustaceans such as some species of shrimp, crab and lobster can remain live out of water for quite sometime if suitable environment is provided in the package and hence can be transported out of water. The important requirements in the packaging for these animals for live transport are :

- maintenance of sufficiently low temperature that will be tolerated by the animals throughout the transit period.
- maintenance of high relative humidity, approximately 70 per cent, in the package so that the animals will be maintained moist.
- availability of sufficient oxygen in the package.

16.3.2 Plastic-bag Method

This is a method most commonly employed for live transport of ornamental aquarium fish and, fry, fingerlings and post-larvae of cultured fishes. Good water quality, healthy fish, proper loading density, and good catching and handling techniques are essential for packaging aquarium fish. For marine species especially, one or two week's conditioning after capture is essential to reduce mortality and also to ensure that they are disease-free. Live corals also may be packed along with the fish; corals stimulate natural conditions.

The fish is placed in a plastic bag partially filled with clean water. This is carried inside a cardboard box insulated with 2.5-5.0 cm thick polystyrene foam. Air is removed from the bag and it is inflated with oxygen. The bags are then sealed and shipped after sealing the outer cardboard box.

16.3.3 Tank Method

This is the method employed in bulk transport of live fish. The method involves shipping live fish in live well boats or tankers. Such units are equipped with circulation and aeration systems. The circulated water is aerated before feeding back to the tank. Water can hold more oxygen in solution at low temperatures; however, fish require more oxygen at higher temperatures. Therefore a tank of a given volume can hold more fish at lower temperatures than it can hold at higher temperatures. That is the reason why the temperature of water in transportation is always kept low according to the levels that the fish species concerned can tolerate. Water temperature is controlled by adding ice and also by insulating the tanks.

One of the factors governing the load of fish in a given volume of water is the species of the fish, because different species of fish have different oxygen demands. An average load of 200-250 kg fish in 1000 litre water is often possible. For short journeys it can be even 450-500 kg fish per 1000 litre water.

16.4 HANDLING OF FEW SPECIES FOR LIVE TRANSPORTATION

Pre-package handling as well as packaging needs will vary with the species of fish/shellfish concerned. Some species of fish and many molluscan shellfish have the capacity to remain live out of water for quite some time if some primary requirements are met. Molluscan shellfish like clams, mussels, oysters, and crustaceans such as some species of shrimp, crab, lobster etc. can be basically handled similarly. Proper handling should begin from the harvest itself.

Most of the common molluscs such as clams, mussels, oysters etc. grow in the littoral and inter-tidal zones. They are filter feeders and the shells will remain agape while under water. While filtering water, they also ingest the bacteria present and hence at any time, may be loaded with bacteria, and consequently, may contain many pathogens. If the water is polluted this possibility is very high. Therefore the molluscs are depurated by holding them in tanks of clean water, where they open and respire, but gets no feed. The animals empty their gut contents including bacteria and thus get cleansed of the bacterial load. The water in which the animals are held should have been made bacteria-free by treatment with ultraviolet light or ozone.

16.4.1 Mussels

Mussels, with their thin walls, are susceptible to easy damage. They should be harvested in clusters to reduce water loss. The byssus gland should not be disturbed; however, the byssus threads many be trimmed. Mussels after harvest should be immediately chilled. Chilling can be done by ice, however, without direct contact with the animals. For this a piece of perforated material such as sacking or cheesecloth may be placed over the animals and ice placed over it allowing for the ice melt water to drain away. Mussels are chilled to 2-4°C by this method. Generally, 10 kg lots are packaged in polybags, sealed and further packed in polystyrene box filled with flake ice and the box is finally sealed with adhesive tape.

16.4.2 Oysters

Oysters should be kept cool in humid atmosphere. After harvest they may be held refrigerated onboard. Alternately they may be kept in holding tanks with provision for recirculating seawater.

Healthy oysters are heavy and will remain tightly closed or shut quickly when disturbed. They should be carried with the cupped half of the shell downward to allow for retention of the liquid within the flat half of the shell. Live oysters that are prechilled to 1-10°C are transported in polystyrene boxes or waxed cartons with inner polythene lining. Oysters are also packaged in sacks, barrels or specially compartmentlised containers depending on demand.

16.4.3 Shrimp

For short distance transport of live shrimp, water is the most common medium used in closed or open systems. For long distance transportation by air, a hibernation technique is used in the case of some species. For species of shrimp like *Penaeus japonicus*, that can stand lower temperatures upto 4°C and survive in a moist atmosphere, transport in chilled moist sawdust is advantageous.

Penaeus japonicus for live transport is collected in aerated tanks and the water temperature is slowly lowered by addition of small blocks of ice. As the temperature of water, which should be continuously monitored, falls to 4°C hibernation of the shrimp begin to take place. A fibreboard box lined with PVC sheets is used as the container. The hibernating shrimp is placed in the box over a layer of chilled sawdust and are immediately covered with another layer of chilled sawdust. Filling with alternate layers of shrimp and sawdust is continued until the box is full. The packaging temperature in the box should be 4-10 °C. Shrimp will remain live under these conditions for more than 14 hours; and hence air transport is possible.

Algae can be used in place of sawdust. Algae have the advantage that they will keep the atmosphere humidified. However, when sawdust is used, it should be ensured that it is free of resins, is untreated and is free from pesticides. Temperature can be maintained for extended periods if insulated boxes are used as containers.

For shrimp species like *P. monodon* or the giant freshwater shrimp *Macrobrachium rosenbergii* a closed bag system or an open tank system using aerators can be conveniently used for relatively shorter journeys. The load density in either case will depend on the journey time involved.

Brood stock and post-larvae of shrimp are transported live employing the closed plastic-bags system; whether for inland transport or for overseas transport by air. The package density of brood stock is 2-10 pieces per 10 litre water. The low package density is to avoid overcrowding which may eventually damage berried females. Fry or post-larvae, depending on their size and distance to be traveled, can be stocked at 1000-2000 numbers per bag.

16.4.4 Lobster

Lobster is transported live under conditions of low temperature, availability of sufficient oxygen, and appreciably high humidity. Lobsters after harvest are maintained in tanks under conditions of minimum stress until packaged and transported.

A popular system of holding lobster in tanks is a closed tank system where seawater is circulated through a biological filter. The filter consists of various grades of shell grit held in layers where bacterial colonies adhere and grow. As the seawater passes through the filter, the resident bacterial flora convert potentially toxic byproducts such as ammonia and nitrite into less toxic substances.

The container used should be light, leak-proof, reasonably strong and easy to handle. Thick-walled polystyrene containers or corrugated fibreboard cartons lined with polystyrene foam will be ideal containers to meet the above needs.

Lobsters should be prechilled in water over a 12 hour period to about 4°C before packing. Only healthy lobsters identified by their quick response to a prod to the antenna between their eyestalks should be packed in the container. The animals are generally held for 24 hours in the holding tanks to identify the healthy ones and sort out the weak and dying. Lobsters are packed in boxes laid with ice pads, and impregnated with silicon to absorb melt water. The prechilled lobsters are packed in layers separated by a moistened material such as seaweed, wood shavings or similar materials to maintain the relative humidity at 70 per cent. A final layer of moistened material is placed on top and the box is sealed with a lid. The temperature is maintained at 1-7°C.

16.4.5 Crab

Crabs should be stored at least 24 hours prior to packaging and shipping so that weak, moribund or injured specimens can be removed. They can be held alive on vessels or in tanks onshore. For deepwater crabs, ideal temperature of holding water is 0-5°C. Warm water crabs can be held at higher temperatures; however, to reduce the metabolic rate and cannibalism, lower temperatures are preferred.

Control of temperature and humidity are the critical factors affecting mortality during shipment. Maintenance of high humidity above 75 per cent and temperature slightly lower than that of the natural environment of the crabs concerned is very essential in transport packages of live crabs.

Use of ice in the package is not considered advisable since very low temperature and contact with ice melt water are lethal to most species of crabs. Therefore, frozen gel packs are used to bring down the temperature. A frozen gel pack is first placed in the container over which moistened material such as coarse canvas or jute, seaweed or wood shavings is spread. The temperature-controlled crabs are then layered over this and the packing is continued with alternate layers of crab and moisture absorbing materials as above. A final layer of moistened material is placed at the top. The insulated lid is then secured in place and sealed. Crabs packed this way will survive for 24-30 hours out of water. At the destination the animals in good condition should be placed in holding tanks.

Mud crab (*Scylla serrata*) can survive out of water for a week, or so, if maintained cool and moist. They can be transported out of water if protected against drying out. However, the claws and legs should be tied to avoid inflicting injury to the neighbours and also the personnel handling them. Tying should not be too tight as this can affect the meat quality.

Suggested Reading

Berka, R. (1980) *The Transport of Live Fish : A Review,* FAI EIFAC Technical Paper No. 48. FAO, Rome.

Fry, F.E.J. and Norris, K.S. (1962) The transportation of live fish. In: *Fish as Food,* Vol. 2., Borgstrom, G. (ed.), Academic Press, New York.

Johnson, S.K. (1979). *Transport of Live Fish;* Publication No. FDDL-514. Fish Diagnostic Laboratory. Agricultural Extension Service. Department of Wildlife and Fisheries. The Texas A & M. University System, College Station.

Rajadurai, P.N.R. (1989) *Live Storage and Distribution of Fish and Shellfish.* Infofish International 3/8.

Rao, C.V. N. (1993) Packaging of Fresh and Live Fish. In: *Fish Packaging Technology, Materials and Methods,* Gopakumar, K.(ed.), Concept Publishing Company, New Delhi.

Wheaton, F.W. and Lawson, T.B. (1985) *Processing of Aquatic Food Products,* John Wiley, Chichester.

17
PACKAGING

PACKAGING is an essential requirement virtually for every product, whether it is a food or any manufactured item. Packaging sells the product; for that it should be delivered to the customer or consumer in the most appropriate type of container. Apart from the selling aspect, packaging has several other vital functions to perform. For food products, packaging is also a means of conserving and preserving it. Packaging should protect all the innate qualities that have been built into the product before packaging. Packaging requirements will vary with respect to the product packaged. Unlike many other manufactured products like consumer goods, leather, machinery etc. the packaging needs of food and food products, and particularly fish, are very complex because of the intrinsic characteristics of such products and the need to retain or preserve them while in the package.

17.1 FUNCTIONS OF PACKAGING

Packaging is an external means of preservation of a product from spoilage, damage or loss due to external causes. Most of the packaged foods are subjected to some form of preservation before packaging. Packaging material may help to further extend the shelf life of the product. A food package can be considered as a structure designed to contain a food in order to

- protect it against contamination, loss or dirty handling.
- protect it against damage or degradation.
- provide a convenient means of dispensing and selling.

- make it easier and safer for transport.

The above functional requirements of packaging imply that the packaging has to perform functions other than protection and preservation alone of foods. Commanding consumer appeal through an aesthetic presentation also should be considered a function of packaging. Basic concepts in a package, in general, are :

- protection and preservation
- containment
- machine performance
- communication
- convenience
- economic considerations

17.1.1 Protection and Preservation

One of the important requirements in a packaging is that it should be able to protect the goods from the outside environment as well as from the rigours of transport. The package should provide an environment suitable for the product to retain its intrinsic qualities when the outside conditions are unfavourable.

17.1.2 Containment

Another requirement is that the package should be able to hold and keep the contents together during transit. The packaging should also be such that it permits easy handling and storage.

17.1.3 Machine Performance

Packaging and package formation are now mainly machine-operated functions. Therefore it is important that the packaging materials should perform satisfactorily on the packaging and package forming machines. The factors influencing the selection of packaging materials with regard to machine performance are

- flexibility.
- printability.
- ease of working on wrapping machines.
- heat sealability.
- suitability for forming by blow-moulding, extrusion, vacuum or thermal techniques.

17.1.4 Communication

In order that the product sells, the package should communicate. This is particularly so for retail packages. The main requirement in communication is the identification of the product contained in the package. Package also may display any other information relevant to the product. In a food product these can be the expiry date, cooking and serving or other instructions as to how to consume, nutrition label and any other statutory information. Non-retail packs also should communicate. Handling instructions, destination, ownership, type of storage etc. are the information to be displayed on such containers.

17.1.5 Convenience

The convenience factor includes the ease with which the package can be opened and the method of its dispensing. The package design should be simple to open and should have no need for any contraption. The disposal of the package also should be easily met with; it should either be reusable or recyclable. Convenience factor should be available in the industrial and transport packages as well. Such packages should be convenient to handle in transit, fit for the mode of transport employed, meet the needs of storage in warehouses and picking up in the best possible way.

17.1.6 Economic Considerations

While there cannot be any compromise on the basic requirements in a package, the package itself should not add highly to the product cost making it exorbitant. The package should be simple and as inexpensive as possible consistent with the primary objectives of protection and attractiveness at sales point.

17.2 SPECIAL NEEDS IN FOOD PACKAGING

While the basic requirements dealt with are common for any packaged product, the food packaging are unique in some of their own requirements. Most of the foods are perishable, some having a useful life of only a day or two. Hence the requirement of food products packaging materials are generally more stringent than for many other commodities. The principal intrinsic requirements of a food packaging material are given below :

17.2.1 Transparency and Surface Gloss for Consumer Appeal

A transparent packaging permits the consumer to see what he is buying and creates confidence about the product in his mind. On the other hand, with an opaque or non-transparent package he has no way of seeing what is in and has to be convinced by his past experience with the product or description by the seller about the quality of the product. The gloss of the package further adds to the consumer appeal.

17.2. Barrier Properties

Barrier properties are mostly related to the water vapour transmission and gas/vapour transmission characteristics of the packaging material. Many products have very low moisture content, and may absorb moisture resulting in spoilage, if the packaging material is inadequate as a barrier against moisture. For such products a package with very low water vapour transmission rate should be employed. It is also necessary to prevent moisture loss from some foods. There are cases where some loss of moisture is desirable to avoid condensation of moisture inside the packages leading to bacterial reactions and mould growth.

In such cases a package with a high water vapour transmission rate will be desirable.

Gas transmission characteristics are also important in food packages. Some fresh food such as fruits and vegetables will respire and hence the packaging material should permit ingress of oxygen and respiration of carbon dioxide. Fresh meat requires ingress of oxygen to maintain the surface colour. However, foods with high fat content such as bacon, fish etc. become rancid on exposure to oxygen. Such foods are packed in an inert or modified atmosphere and hence demand the use of packaging materials with good barrier properties against gas transmission.

Some of the other important attributes expected out of a food package are :

- ability to withstand wide temperature ranges in storage and while in use
- absence of toxic substances. Toxic materials, if present, may migrate to the food and make it toxic
- sturdy enough to protect the food against crushing
- low cost while satisfying the above needs

17.3 PACKAGING MATERIALS COMMONLY USED

Packaging materials widely used for food products are:

- paper and coated-paper products.
- cellulose products - cellophane.
- metals like tinplate, aluminium.
- ceramics.
- glass.
- rubber.
- plastic.
- wood, fibre etc.

Practically all these materials have long been in use for packaging different products providing satisfactory services. However, plastics which are of relatively recent origin, while offering several advantages over other packaging materials, have been presenting several problems as well for food products. Some of the chemical adjuvants used in the manufacture of plastic materials may be toxic in nature and may get transferred to the food when the package is in contact with the food material. However, plastics have the advantage that most of them possess excellent physical properties such as strength and toughness, simultaneously being light in weight and flexible, as well as resistant to cracking. A wide range of polymers is now available for conversion into diverse types of plastic packaging materials. Often the requirements with a particular food may not be met with a single material, as it may not possess all the desired properties. In such cases copolymers or laminates consisting of two or more layers of different polymers having different properties can be used.

17.3.1 Plastic Packaging Materials

The polymers most commonly employed in packaging food products are :

- low density polyethylene (LDPE).
- high density polyethylene (HDPE).
- polypropylene (PP).
- polyvinyl chloride (PVC).
- polyvinylidene chloride (PVDC).
- polytetrafluoroethylene (PTFE).
- polyamides (Nylons).
- polycarbonates.
- polyesters.
- rubber hydrochloride.

Most of the polymers used in the manufacture of food packaging are thermoplastics because they have to be subjected to several heating and cooling cycles during synthesis and conversion into films or formed containers. Thermoplastics have the advantage that they can be repeatedly softened and hardened by heating and cooling respectively, provided no chemical decomposition takes place.

17.3.1.1 *Ethylenic Thermoplastics*

Most of the thermoplastics are derivatives of ethylene. They are also known as vinyl plastic or polyolefins. The important among the ethylenic polymers used in plastics are

• POLYETHYLENE (PE)

Polyethylene (polythene, PE) is the material consumed in the largest quantity by the packaging industry. It is made by polymerisation of ethylene, which in its natural state is a gas. Polymerisation under very high pressure (1200 atm.) and high temperature (150-200°C) in presence of traces of oxygen yields a macromolecule with very high degree of branching, whereas at low-pressure forms polymers with long parallel linear chains. Because of the branched chain structure, close packing of the polymer chains is not possible. Therefore the density of the finished product is low and the product is called low-density polythene (LDPE).

LDPE is a strong, slightly translucent flexible material possessing excellent resistance to most chemicals below 60°C. LDPE is soluble in hydrocarbon and chlorinated hydrocarbon solvents. It is a good barrier against water vapour, but not so good against oxygen and other gases. It cannot be employed in a packaging likely to be subjected to heat treatment because of its low softening point. It is amenable to fusion welding to itself to give excellent liquid tight seals. However, it can stand low temperatures. It is also subject to stress cracking in presence of polar chemicals.

LDPE can be readily converted into lightweight films with good heat sealability. It can be coated onto other materials such as aluminium and paper.

High-density polythene (HDPE) is produced using a low temperature (60-160°C) and low pressure process employing alkyl metal catalysts in the process. It is stiffer, harder and less transparent, but possesses greater resistance to oils and greases. It offers excellent moisture protection and has a higher softening point, but the impact strength is low.

• POLYPROPYLENE

Polypropylene (PP) is a product of polymerisation of propylene. Under controlled process of polymerisation employing stereospecific catalysts and low pressure it is possible to produce a polymer in which the methyl groups are largely arranged on the same side of the polymer chain. The resultant polymer will have higher crystallinity. However, this will be less than that of HDPE. Other important properties of PP are :

- more rigid, stronger and lighter than PE.
- harder, and higher softening point than HDPE.
- greater resilience, good gloss.
- greater resistance to chemicals except hot aromatic and chlorinated hydrocarbon solvents.
- low water vapour permeability.
- good grease resistance.
- lower shock resistance at lower temperatures.
- stable at high temperatures.
- can take on print without difficulty.

Polypropylene is generally used in the oriented forms. Biaxially-oriented PP films, i.e., the films stretched in two directions at right angle under suitable temperature conditions, will have improved gloss and clarity, tensile strength, abrasion, and barrier properties against water vapour and oxygen.

• POLYVINYL CHLORIDE (PVC)

This is a polymer of vinyl chloride (CH_2CHCl) i.e., ethylene in which one hydrogen atom is substituted by a chlorine atom. PVC has low crystallinity, and is polar in character. It is a hard, stiff, clear glossy material with excellent moisture resistance and low gas permeability. It is resistant to most solvents and oils and is less flammable. By addition of palsticisers flexible films can be obtained. The film is non-toxic.

• VINYLIDENE CHLORIDE (VDC)

In this polymer two hydrogen atoms of ethylene are substituted by chlorine atoms, CH_2CCl_2. It is a hard solid with high degree of crystallinity, is insoluble in most solvents and has very low water absorption. The film produced is clear with very low permeability to vapours.

• POLYVINYLIDINE CHLORIDE (PVDC)

PVDC is usually produced as a copolymer with 13-20 per cent vinyl chloride. It contains a low per cent of plasticisers, slip agents, and stabilisers. These films are clear, have excellent mechanical resistance and extremely low water vapour and gas transmission rates.

• POLYTETRAFLUOROETHYLENE (PTFE)

This is a polymer of ethylene of which all the hydrogen atoms are substituted by fluorine atoms, C_2F_4. It is highly crystalline, has high molecular weight and is smooth and waxy. It can produce a non-stick surface and withstand high temperatures.

• POLYSTYRENE

Styrene is ethylene in which one hydrogen atom is substituted by a phenyl radical, CH_2-CH-C_6H_5. In the polymerised product, phenyl radical prevents close packing of the macromolecular chains. Therefore it is amorphous and brittle in nature. It is transparent with a high refractive index, and is a good barrier against gases, but not against water vapour. It is soluble in a number of solvents.

17.3.1.2 Non-ethylenic Thermoplasts

• POLYAMIDES (NYLONS)

Polyamides are condensation products of diacids and diamine. The first polyamide produced was Nylon-6,6 made from adipic acid and hexamethylene diamine. Various grades of nylons are available. Nylon-6 is easy to handle and is abrasion-resistant. Nylon-11 and nylon-12 have superior barrier properties against oxygen and water and have lower heat seal temperatures. However, nylon-66 has a high melting point and hence it is difficult to heat seal. Nylons are strong, tough, highly crystalline materials with high melting and softening points. High abrasion resistance and low gas permeability are other characteristic properties.

• POLYCARBONATES

Polycarbonates are linear polyesters of carbonic acid with aliphatic or aromatic dihydroxy compounds. The films are heat sealable, and have excellent clarity. They have high impact strength and good temperature resistance. They are resistant to weak acids and alkalis, and the permeability to gases and water vapour is poor. These properties are maintained over a wide range of temperature depending on molecular weight. These compounds are readily soluble in chlorinated hydrocarbon solvents.

• RUBBER HYDROCHLORIDE

Rubber hydrochloride is produced from natural rubber by treatment with hydrochloric acid. It can be stretched, is heat sealable, non-toxic, resistant to oils and greases, unaffected by acids or alkalis and, does not support combinations. On storage it may become discoloured and brittle.

17.3.2 Copolymers

The polymers discussed above are all made from a single monomer each and are known as homopolymers. In some cases polymers are prepared by the additive reaction of more than one monomer or by condensation of different monomers. These products are known as copolymers. The properties will differ from those of the individual homopolymers. Their properties will depend not only on the composition of the monomers used, but also on their arrangement in the chain. The important copolymers in food packaging applications are Ethylene-Vinyl Acetate, Vinyl Chloride copolymers and Polyester copolymers

• ETHYLENE VINYL ACETATE (EVA)

EVA copolymers are made by copolymerisation of LDPE with vinyl acetate. They contain up to 20 per cent vinyl acetate and hence are similar to LDPE in properties with some improvement in transparency. EVA's are more flexible than LDPE, but have higher permeability to gases and water vapour. EVA's are thermally unstable at high temperatures, but are very stable at lower temperatures, and have wide heat seal range.

• VINYL CHLORIDE COPOLYMERS

Vinyl chloride is often copolymerised with vinyl acetate, polypropylene and acrylonitrile. Vinyl chloride copolymers have better flow properties simultaneously remaining tough. Vinyl chloride/vinylidine chloride copolymers have low water vapour and gas permeability, and are used to improve the barrier properties of paper, polypropylene and cellulose films.

• POLYPROPYLENE COPOLYMERS

In order to overcome the poor cold temperature impact strength of polypropylene, ethylene is introduced into the propylene structure. Propylene-ethylene copolymers are hazy due to light scattering. Polypropylene copolymers have lower melting points and have improved low temperature impact strength.

• POLYSTYRENE COPOLYMERS

Copolymers of polystyrene, acrylonitrile and butadiene are rigid and possess high impact properties. High nitrile resins have excellent gas barrier properties as well.

17.3.3 Ionomers

Ionomers are essentially LDPE, wherein ionic forces due to carbonyl ions containing metallic ions such as sodium or zinc are introduced. Compared to LDPE, ionomers are more resistant to oils and fats, superior in transparency, toughness, abrasion resistance and coating addition. They are specially used in skin packaging, meat packaging and in combination with polyamides for packaging oils and fats.

17.3.4 Laminates

The packaging needs of food industry is so diverse that no ideal universal packaging is available nor is it practicable to develop one. None of the polymers discussed above possess all the desirable properties for all food products. So, the packaging requirements are often met with employing several polymers or using other types of materials like aluminium or paper or their combinations. Such combinations are called laminates and consist of layers of individual materials on top each other, and are bonded together with the help of adhesives. Laminates containing all plastic materials can also be produced as a composite film by co-extrusion or coating.

17.3.5 Other Constituents of Plastics

Plastics are high molecular weight products and do not get transferred to the food they are in contact with. However, some substances are added during the manufacture of plastics to assist the production process or to enhance the mechanical and other properties, and stability of the final product. These include anti-oxidants, anti-block agents, heat and light stabilisers, plasticisers, lubricants, pigments, fillers, fungicides etc. These may be present in varying quantities in the final product and may migrate to the food in contact with them. The possible toxicity of such low molecular weight products is of great concern with a food packaging material.

Another source of substances other than the basic

polymer in a plastic is the polymerisation residues, including monomers, catalysts, solvents etc. The volatile gaseous monomers like ethylene will be lost during storage, but very low levels may persist, and some others like styrene are difficult to remove. Chemicals added as aid in processing are likely to be present in greater amounts and hence these should be selected only from those approved for food contact applications.

17.3.5.1 Migration of Constituents

Migration is a matter of some consequence in food packaging. Migration in the food-packaging applications involves transfer of substances from the package, generally a plastic material, to the food contained by it. Substances that are transferred as a result of contact with the food are generally the monomers and processing additives or the residues of polymerisation. Migration of some constituents of the food to the plastic also can take place. However, it is the migration of constituents of the packaging to the food that is of major concern since many of the migrants may be toxic to humans.

17.3.6 Toxicological Aspects of Food Packaging

Toxicity in foods caused by the packaging materials results from the migration of toxic substances while in contact with the food. Some of the chemical substances added to the packaging for attaining better functional properties may be toxic to humans. Toxicity may be induced by some of the monomers or polymerisation residues as well. The toxic effects of some of the monomers used in plastics manifest in several forms.

Vinyl chloride monomers have been associated with lesions of the bones in the terminal joints of the fingers and toes as well as changes in the liver and spleen of human beings. Long-term exposure leads to a rare form of liver cancer. Vinylidine chloride has been found to affect the activity of several rat liver enzymes and decreases the store of glutathione. Acrylonitrile is more toxic than chlorinated monomers and is associated with carcinogenicity in animals and humans. Styrene and its oxide have been incriminated against chromosomal aberrations in certain conditions.

Even though toxicity and carcinogenicity associated with monomers are fairly well established, no such evidence is available against the polymers themselves. However, with improvements in technology the presence of monomers is getting consistently reduced in the final product, and such hazards are not of great significance with packaging in the polymer.

17.4 PACKAGING NEEDS FOR FISH AND FISH PRODUCTS

Some of the early packaging for fish consisted mainly of wooden crates, woven bamboo/palm baskets, wooden pickling barrels and the like. In tune with the commercial requirements and consumer preferences, rational changes have been brought about in fish packaging as well. Packaging materials made of metal, glass, paper, and particularly plastics have become very popular.

Fish is sold fresh, or iced and transported to distant locations for sale as fresh fish, and is also processed into various products for retail and wholesale. Depending upon the intrinsic qualities of the fish used and the nature of the product processed out of them there are bound to be differences in the packaging needs. Plastics is the main material used for many purposes in packaging fish and fish products. The properties and limitations of individual packaging plastic materials have been dealt with in the preceding paragraphs. Characteristics of each product going into the package should be well understood and a packaging material or combinations to meet the identified needs has to be developed.

17.4.1 Packaging Requirements of Individual Products

17.4.1.1 Fresh/Iced Fish

Fish after catch will remain fresh only for a limited period, 4-6 hours, depending on the environmental conditions and the intrinsic nature of the fish. Fish is normally iced and sold in local markets or transported to nearby locations for marketing. Fish sold immediately in local markets may not need any special packaging. But a proper packaging to ensure better shelf life becomes essential when it has to be transported to distant localities for disposal. A suitable package for fresh fish shall

- provide barrier against oxygen to reduce fat oxidation.
- keep the fish moist and prevent dehydration.
- retard chemical and bacterial spoilage.
- prevent permeation of external odours.

For bulk packaging, the container.

- shall be sturdy enough to withstand the rigours of transit and travel by different modes.
- shall be of light weight, hygienic and easily cleanable.
- shall possess good insulation properties.
- shall have good barrier properties.
- may be returnable or non-returnable on economic considerations.

The following packaging materials can be considered suitable in such circumstances :

- whenever the sales point involves a journey of less than 16 hours from the despatch site, non-returnable inexpensive containers like bamboo baskets can be ideally used for transport of iced fish.
- when the journey involves 16-60 hours, corrugated polypropylene boxes/rigid HDPE boxes will be convenient. Corrugated boxes are cheaper than wooden boxes. They have better insulation properties, and are significantly lighter in weight. For further protection of fish, inner insulation also may be used.

Foamed polystyrene/polyurethane slabs will provide good insulation properties.

When the fish is sold in retail under refrigeration through supermarkets, an ideal and most popular type of container is a shallow, thermoformed tray of polystyrene with a transparent film over-wrap. The tray may be made of semi-moisture proof cellophane, polyvinyl chloride etc. However, such packages may also need an absorbent blotter to absorb any water drip from the product which, otherwise, will make the product messy in appearance.

17.4.1.2 Frozen Fish

Though several frozen fish and fish based products are sold in retail in the domestic markets, frozen fish and fish products in India are identified mostly as an item of export. Though there cannot be any compromise on the packaging requirements *vis-a-vis* maintenance of quality of the products in the domestic markets, export packaging have to adhere to the international specifications and hence are more stringent in their requirements.

The factors affecting the quality of frozen fish are :

- moisture loss.
- oxidation.
- rancidity.
- change in odour and flavour.
- loss of volatile flavours.
- enzymic activity.

The important requirements in the package for frozen fish are :

- flexible enough to fit the contours of the fish leaving little or no free space.
- shall not become brittle at the cold temperature of storage.
- shall not deteriorate in cold storage.
- shall be resistant to puncturing.

- shall prevent or reduce moisture loss.
- shall be impervious to oxygen and gases.
- shall be easy to fill.

One of the most essential requirements in a package for frozen fish is its barrier property against oxygen and water vapour, since these are two factors which significantly affect the organoleptic quality of frozen fish during storage. The most popular packaging in use is made of polyethylene with or without a waxed paper over-wrap. Such packing can take care of most of the requirements of package mentioned above.

The bulk retail packs intended for home freezers are mainly packed in polythene-coated paper bags. These bags, as the contents are used, can be rolled down and re-closed by hand.

The export packaging of frozen fish is of a different type because of several other factors associated with bulk packaging, shipping and trade requirements.

Frozen blocks of fish are given an over-wrap of polythene (this can be film bags in the case of individually quick frozen shrimp or fish), and are then packed in duplexboard cartons. A definite number of unit cartons are packed in a master carton which serves as shipping container. Export package for frozen fish/shrimp has the following components :

Inner wrap	-	LDPE or high molecular weight HDPE or linear LDPE film
Unit carton	-	Waxed duplex cartons
Film bags	-	LDPE or high molecular weight HDPE or linear LDPE
Master cartons	-	Waxed corrugated fibre board cartons with high bursting strength, puncture resistance and compression strength

17.4.1.3 Dried Fish

Moisture and oxygen are principally responsible for spoilage in dried fish. Dried fish may be fragile and is subject to contamination by foreign flavours. They are also subject to insect attack. Carotenoid pigments in products like dried shrimp are sensitive to light; light may also accelerate development of rancidity. Moisture absorption will lead to accelerated spoilage due to bacterial action. Access to oxygen will accelerate oxidative rancidity and the resultant flavour changes. Mechanical abrasion may cause the products to undergo physical disintegration and this is very significant in the case of freeze-dried products. Therefore, the packaging material selected for dried fish shall :

- prevent gain or loss of moisture.
- be impermeable to volatile flavouring compounds.
- have good barrier properties against oxygen.
- be resistant to mechanical abrasion and puncture.
- be impermeable to light.
- be inert.

Polyethylene/polypropylene packaging films commonly used for retail packing have good tearing and bursting strengths, but their water vapour and gas transmission rates are high. They are also prone to puncture or damage from sharp spines of the dried fish. Hence shelf life becomes limited. The packaging materials which can be advantageously used for dried fish products are :

- polyester laminated with LDPE
- polyamide laminated with polyethylene

For freeze-dried products, packaging in paper/aluminium foil/polyethylene laminates or in tin containers under inert gas is ideal.

17.4.1.4 Salted/Smoked Fish

Salted and smoked fish are products with improved/modified flavour. Depending upon the methods employed in processing, these products will have different moisture contents. Smoked fish, however, is preferred mainly for its smoke flavour. Salted/smoked fish can also be frozen and cold stored. In smoked fish rancidity may develop while in frozen storage.

Polyamide polyethylene laminates and polyester polystyrene laminated films are recommended for salted/ smoked fish also. However, greater care is needed in handling and packaging of smoked fish where anerobic packaging has been known to cause food poisoning by the toxin produced by *Clostridium botulinum* Type E.

17.4.1.5 Canned Fish

A suitable packaging for thermally processed fish should have the following properties :

- hermetically sealable.
- thermally conductive.
- shall withstand heat processing at high pressure and temperature.
- shall not affect colour, odour, flavour, texture, and food value of the product.
- be inert to the contents.

Several materials conforming to the above requirements of canned foods are available and are widely in use.

- tinplate cans with sulphur-resistant lacquer coating on the food contact side.
- aluminium cans made of aluminium alloyed with magnesium/manganese and inside coated with lacquer.
- cans made of tin-free steel.
- retortable pouches of 3-ply material, polyester/ aluminium foil/polypropylene.

17.4.1.6 Fish Pickle

Fish pickle contains fish/shellfish, several spices, vinegar and oil. It is spicy, oily and contains appreciable quantities of water. It is used as an appetiser in many Indian diets. Because of these unique features of the pickle, it is often packed in glass bottles. The recent trends, however, is packing fish pickles in flexible plastic packaging. A packaging material for fish pickle shall satisfy the following requirements.

- shall be grease-proof.
- shall have only low water vapour transmission rate.
- shall be impermeable to volatile spice flavour.
- shall be inert.
- shall have good heat sealability.

Ideal packaging materials for fish pickles are :

- polyester laminated with LDPE/HDPE co-extruded film with HDPE contact inside or Nylon-surylyn co-extruded film or LD/Nylon/primacore co-extruded film.
- wide mouthed glass or polyester bottle.

17.4.1.7 Battered and Breaded Products

Several value added products such as battered and breaded fish/shellfish are now popular the world over. Battered and breaded shrimp in different styles, squid rings, scallop etc. are very popular in the restaurant trade. Fish fingers, cutlets, burgers and the like are popular in the fast food trade. These are all products commanding high prices and this will naturally reflect in their packaging needs also. All these products are stored frozen and some of the problems encountered during storage are desiccation, discolouration and development of oxidative rancidity. Therefore the packaging used shall principally have :

- only low water vapour transmission rate.
- low oxygen permeability.
- low permeability to volatile flavours.
- sufficient mechanical strength to withstand the hazards of transport.

The packaging materials identified for such products are :

- thermoformed trays of PVC with transparent film over-wrap.
- thermoformed trays of polystyrene with transparent film over-wrap.

17.4.1.8 Fish Curry

Fish curry is a product presented in a 'ready to serve' style which can be preserved by freezing or heat processing. The problems encountered with frozen and stored fish curry are :

- desiccation.
- discolouration.
- development of rancid flavour.

When heat processed in conventional tinplate or aluminium cans, the product becomes downgraded due to development of a metallic taste and a certain extent of discolouration. The appropriate containers that can be used for fish curry are :

Frozen curry	- Thermoformed trays made of polystyrene or polyvinyl chloride with a transparent film over-wrap
Canned curry	- Retortable pouches made of polyester/ aluminium foil/cast polypropylene

17.5 ENVIRONMENTAL IMPACT AND LEGISLATION

Because of the possibility of some of the constituents of the packaging materials migrating to the food while in contact with it and their adverse impact on human health, several legislative measures have been adopted in several countries regulating the use of plastic containers. Plastic materials are non-degradable and hence their accumulation can cause severe environmental degradation by polluting air, ground water and even the sea. Therefore, the environmental legislation on packaging are mainly based on 'source reduction' and also based on the principles of 'reuse' and 'recycling'. Source reduction indicates use of less raw materials, especially of non-renewable type, and material reduction using lighter and smaller packages. 'Reuse' aims at packaging designed to withstand repeated distribution trips, that is the potential for using again and again. 'Recycling' involves the concept that packaging materials after use can be recovered and reused either for packaging or other industrial applications. As source reduction requires long range marketing strategies and considerable changes in packaging designs, the emphasis now is on reuse and recycling. There are several regulatory agencies working in different states to regulate waste disposal and to set up standards for mode of disposal.

Suggested Reading

Gopakumar, K. (ed.) (1993) *Fish Packaging Technology, Materials and Methods.* Concept Publishing Co., New Delhi.

Griffin, R.C. and Sacharow, S. (1972) *Principles of Package Development.* AVI Publishing Company, Westport, Conncecticut, USA.

Paine, F.A. (ed.) (1987) *Modern Processing, Packaging and Distribution Systems for Food.* Blackie Academic & Professional, Glasgow and London.

Paine, F. A. and Paine, H.Y. (eds) (1983) *A Handbook on Food Packaging.* Blackie Academic & Professional, London.

Sacharow, S. and Griffin, R.C. (1970) *Food Packaging.* The AVI Publishing Co., Westport, Connecticut, USA.

18
QUALITY ASSURANCE

QUALITY assurance should necessarily form an integral part of any manufacturing industry. It is very important in the food processing industry, and more so in the fish processing industry because of the highly perishable nature of fish and the probability of many hazards, particularly pathogenic bacteria, pesticide and heavy metals residues etc. generally found associated with fish. A processed fish product reaching the consumer shall be in a style demanded by him, be hygienically prepared, and free from hazardous bacteria and toxin producing substances. Quality of a food product is different in concept when compared with any other article of commerce, because in foods, particularly seafood, quality is influenced by intrinsic as well as extrinsic factors. Intrinsic qualities like chemical composition, nutritive value, species, size, sex, presence or absence of parasites, poisonous nature, contamination with pollutants etc. are inherent in the material and hence are beyond the control of the processor. Extrinsic quality is the sum of the effects of all the treatments the fish receives from the time of catch till it reaches the consumer. Extrinsic quality can be controlled and may include degree of freshness, damage, and deterioration during processing, storage, distribution and preservation for sale.

Quality control of seafood differs very much from that of any other food product. Unlike fruits, vegetables or meat that can be reared to the desired level of growth and harvested in a predetermined place according to predetermined schedules, fish is harvested from the sea from an unknown stock. Therefore, there will be great variations in their intrinsic qualities. The unpredictability of the catch enforces several inconveniences on the processing schedules. Further, tropical fishery is a multi-species one with very small to very large size fishes in the catch, most often each variety only in limited quantities. This is another hurdle making the quality assurance programme in seafood more complicated.

18.1 QUALITY

Quality literally means degree or grade of excellence of a thing/a distinctive attribute or faculty/a characteristic trait. Some of the older definitions like 'fitness for use', 'value for money', 'giving the customer what he wants', 'at the right cost' etc. have undergone gradual changes. The definition widely accepted now is 'the totality of features and characteristics of a product or service which bear on its ability to satisfy the stated or implied need'. It implies that quality is a measurement of whether a product or service provides the features which a consumer considers necessary. It can also be inferred that quality need not necessarily be related to cost, but lies mainly with the satisfaction of the consumer in meeting his desired needs. It also becomes apparent that quality is a concept which is dependent on the consumer's needs. If a product satisfies the needs of a consumer and is within his means, will be a quality product for him; for another consumer who expects more features or conveniences, it will be a product lacking in quality. Hence quality, in a wider sense, can be considered to be all those attributes which a consumer considers should be present in a product.

18.2 QUALITY CONTROL *VS.* QUALITY ASSURANCE

Quality control and quality assurance are two features which should be properly understood to work out a schedule of operation and practices for processing a product of desired quality and should be understood in the proper perspective by the seafood processing personnel.

18.2.1 Quality Control

Quality control is a system of maintaining standards in the manufactured products by testing a sample of the output against the specifications. Quality control of seafood consists of steps taken between harvesting and retail which protect the quality of the final product. It can be considered as the sum total of all procedures and activities utilised in a processing plant to ensure production of quality products. Quality control is vested with three sets of responsibilities viz., quality assurance, quality inspection and quality verification. Quality control, in general, represents the various measures employed to assess whether the procedures and processes designed for a particular product are strictly followed in operation. However, it is different from quality inspection which consists of the monitoring necessary tc measure the effectiveness of quality control programme.

18.2.2 Quality Assurance

The term 'quality assurance' refers to the functional responsibilities associated with the planning and design activities necessary to develop an effective quality control programme. It has the objective of planning a product and the related control measures so that it conforms to the established standards or specifications, and that the product is upto the expectations of the consumer. Quality assurance systems are designed to prevent problems before they occur. A quality assurance programme consists of :

- quality standards.
- quality control/quality evaluation.
- quality audit and inspection.

Aim of a quality assurance programme is to ensure that everyone in a processing chain from harvesting to retail, is working to achieve a high quality end product (see also Section 18.4.1).

18.3 QUALITY ASSURANCE PROGRAMME IN SEAFOOD PROCESSING

A properly designed and implemented quality assurance system will be an effective tool to judge the effectiveness of the entire schedule in the seafood processing operation. The programmes implemented should be sufficient to ensure maintenance of quality, gathering information on the effectiveness of operational steps and meeting the national and international regulatory requirements. Further, there is a greater awareness among the consumers that quality is about the organisation and not with the product, signifying that only a well run establishment with an in-house quality assurance regime can produce consistently safe, wholesome and standard products

18.3.1 Maintenance of Quality

The quality programme implemented has a primary responsibility that it is instrumental in maintaining consistent quality products which satisfy the demands of the customer. The programme should be such that it makes the process economical to the producer and does not impose any additional financial burden on him.

18.3.2 Gathering Information

Information and data on the following aspects are essential to assess the efficiency and effectiveness of each unit operation :

- quality level of incoming raw materials.
- defect level at each unit operation.
- sanitation level in the whole plant as reflected in the microbiological indicators.
- assessment of unit operations for conformity within acceptable levels.
- history of each batch of product from the plant.

The information gathered, besides deciding the adequateness of different operations in processing a quality product, can also be used for increasing the productivity as well as the profitability of the operations.

18.3.3 Meeting National and International Regulatory Requirements

An effective quality assurance programme should ensure that the products conform to the various regulatory requirements in force. Food laws may be different in different countries and the product need not necessarily be consumed in the country of production only. Therefore, these requirements with respect to the products may be subject to the regulatory requirements in force in the country of production as well as of those countries to which the products are proposed to be sold.

In short, a properly planned and implemented in-house quality assurance programme should be able to ensure consistency, competitiveness and quality of the products.

18.4 FUNCTIONS OF QUALITY PROGRAMMES

A processing establishment should have a well-defined quality programme, strict observance of which should lead to the production of quality products. A quality department should be an integral part of the processing establishment to oversee and implement the programme. The functional responsibilities of a quality programme will fall under three heads

- quality assurance.
- quality inspection.
- quality verification.

18.4.1 Quality Assurance

Functional requirements of quality assurance programmes are two-fold. One is that it should establish and maintain the standards for quality of each individual product/process. To achieve this the quality assurance programme should

- formulate specifications for raw materials, supplies, in-plant processes, containers and finished products including their shelf life.
- develop test procedures and testing of quality levels and production variables.
- develop sampling procedures and the frequency of sampling for testing.
- prepare appropriate proforma for recording, reporting and, quality control charts.
- attend to troubles and rectify defects.
- attend the special problems regarding quality and complaints about the products.
- train the required personnel.

The second aspect is it should ensure that requirements, national and international, with respect to the product quality are scrupulously met with.

18.4.2 Quality Inspection

Quality inspection is an in-plant process and has the responsibility to oversee and ensure that the specifications for each process and operation are strictly adhered to at all stages in the processing schedule. Strict maintenance of quality inspection with respect to sanitation and processing techniques can obviate the occurrence of any defect in the final product. Non-compliance with the specification may lead to product failure.

18.4.3 Quality Verification

Quality verification is the functional area which

- analyses whether the process has been able to produce the products conforming to the quality requirements
- assesses the adequacy of quality programmes
- verifies whether the quality specifications have been properly followed in each unit operation
- provides the laboratory support in the process evaluation.

18.5 END PRODUCT QUALITY *VS.* PROCESS CONTROL

Quality control in the beginning relied solely on the end product inspection. It is well understood that the quality of the end product is influenced by several factors right from the quality of the raw material to start with. Even with the best possible raw material, the end product quality may differ when processed in two different establishments one of which maintains a strict reliance on quality assurance programmes and on-line process control, and the other which is rather slack in these respects. In the latter case, the product may often not conform to the quality requirements, whether internal, national or international, and be not acceptable to the consumer. The options before the processor in such a contingency are

- reprocess the product. This may further downgrade the quality of the product due to loss of freshness. Further, increased labour cost, additional cost of packaging and decreased throughput will make the entire operation uneconomic.
- discard the product.
- ignore the customer complaint to the detriment of the future of the processor.

It is obvious from the above that end product inspection to judge the quality is not a reliable method. An alternate to end product inspection for quality is to control the quality of the process rather than the product, that is the process control.

18.5.1 Process Control

In process control, the emphasis shifts from the end product quality to applying a strict code of practice for processing. The emphasis here is on ensuring proper design of products and processes and the utilisation of a quality programme that guarantees that the actual processing operations conform to the design so that the quality of the end product is assured. Product and process designs, and documentation to ensure that all the requirements for achieving the production of quality product and conformance to all quality requirements are met with in each unit operation are the important considerations in this system.

18.5.2 Safety of Foods

Fish processor has the responsibility of providing safe, wholesome and quality product to the consumer. Safety is an essential and integral part in the design of products and processes. Absence or lack of consideration to food safety at these stages may pose serious threats to public health. A food safety management system that is now in force in many countries to ensure that all unit operations of a process are so controlled as to preclude any health hazard to the consumer is Hazard Analysis Critical Control Points (HACCP) and is described in some detail below.

18.6 HAZARD ANALYSIS CRITICAL CONTROL POINTS (HACCP)

HACCP is a total quality management system with emphasis on safety based on a systematic approach to identification, assessment and control of hazards. It is a preventive control system in which hazard is controlled or eliminated before it occurs. It concentrates on prevention strategies on known hazards and the risks arising out of them occurring at specific points in the processing schedule.

18.6.1 HACCP Concept

Food must be safe to consume and conform to certain standards. If some properties are monitored by the plant, but without supplementary information, the tests will provide only a poor means for controlling and operation. If the product does not conform to specifications, it may have to be reprocessed or discarded. This contingency can be avoided if certain key variables in the process are monitored and controlled. Such a system is provided by HACCP. HACCP is based on a set of seven principles. The system envisages identification of potential hazards in seafood processing at all stages upto the point of consumption. The seven principles are

- Hazard analysis - Assess the hazards associated with capture, storage, raw materials and ingredients, pre-process and process operations, and all other activities upto consumption. Prepare a flow diagram of the steps in the process. Identify and list hazards and specify control measures.
- Determination of critical control points (CCP)
- Specification of criteria - Establish target levels and critical limits that must be met to ensure that each CCP is under control.
- Establishment of procedure and monitoring system to ensure control of the CCP and their implementation.
- Corrective action when the monitor indicates any deviation from the critical limits or that the process is out of control.
- Establishment of procedures to verify that the HACCP system is working correctly and effectively.
- Establishment of documents concerning all procedures and records appropriate to these principles and their application.

18.6.2 HACCP Team

In order to fully understand the process and to identify the hazards and critical control points a HACCP team made up of people from a wide range of disciplines should be organised first. It should have the following composition and responsibilities

Chairman	-	Convene the group, direct the work of the team, ensure that the concept is properly applied
Process expert	-	Should possess detailed knowledge about the production

		process, require to draw up the initial flow diagram
Engineer	-	Should have good knowledge and understanding of the mechanical operations and performances of the processing stages
Microbiologist Hygiene specialist Food technologist QC manager	-	Specialists with understanding of particular hazards and associated risks
Technical Secretary	-	Recording of team's progress, results of analysis etc.

Persons such as packaging specialists, raw material buyers or distribution staff may be brought into the scheme temporarily in order to provide relevant expertise whenever needed.

18.7 STEPS IN THE HACCP STUDY AND EXPLANATION OF TERMS

18.7.1 Hazard

A hazard is a biological, chemical or physical factor that has the potential to cause an adverse effect on human health. The HACCP team should list all hazards that can be reasonably expected to occur at each step.

Biological hazards include pathogenic microorganisms, parasites, toxigenic plants, animals and products of decomposition like histamine. Pesticides, detergents, antibiotics, heavy metals, non-permitted food colours and food additives etc. constitute the important chemical hazards. Extraneous matter like filth, metal or glass fragments, stones etc. are important among the physical hazards.

18.7.2 Hazard Analysis

This is a system using which the significance of a hazard to consumer safety can be analysed. By using this system it can be decided which hazards are of such nature that their elimination or reduction to acceptable levels is essential to produce a safe food product. Identification of hazards, their assessment, and identification of control measures constitute the important functions of hazard analysis.

18.7.2.1 Identification of Hazards

Any food-borne illness identified as associated with consumption of a particular product is evidence enough to suggest the existence of an uncontrolled hazard. The concerned food should be subjected to detailed analysis to ascertain the cause of the illness. Once this is done the stage at which the food might have contracted it should be ascertained. Epidemiological records should be able to give information on the known factors responsible for the outbreak of food-borne illness for particular products. In the absence of epidemiological information, technical information can be gathered on all aspects of production, processing, storage, distribution and use of the product. This should include the hygienic design of equipment, hygiene and sanitation procedures in the plant and, personnel health and hygiene.

18.7.2.2 Assessment of Hazard

A quantitative assessment of hazard with due regard to risk and severity of the identified hazard should be carried out. Risk is the chance of a hazard occurring and severity, its magnitude. An 'unacceptable health risk' means a risk that exceeds the regulatory action levels, tolerance or other limits established for the hazard. It is the responsibility of the HACCP team to decide which hazards are significant and must be addressed in the HACCP plan.

18.7.2.3 Identification of Control Measures

After assessing the hazard, the HACCP team should consider which of the existing control measures could be applied for each hazard. Often, more than one control measure may be needed to control a specified hazard; more than one hazard may be controlled by a specific control measure as well.

18.7.3 Critical Control Points

A critical control point (CCP) is a point or stage in the processing operation where failure to control effectively would most likely result in the production of defective/ unsafe food. In other words, it is a step which, if properly controlled, will eliminate or reduce a hazard to an acceptable level. A step/point is any stage in the production. This includes raw materials, transport to processing plants, processing and storage. CCPs require constant checking to ensure compliance with all the requirements of the product.

18.7.4 Control Points

These are other points in the processing operation where failure to effectively control may not necessarily result in the production of defective/unsafe food. These points require occasional checking throughout the production shift.

18.7.5 Determination of CCPs

The relevance of each identified hazard in each stage of the process is considered. If the hazard can be reduced, prevented or eliminated through some form of control at a particular stage, it is a CCP.

There are two types of CCPs, CCP1 and CCP2. CCP1 will assure the control of a hazard, whereas CCP2 will minimise the hazard, but will not assure its control. Both are important and should be controlled. A judgement of risk also must be made so that a level of concern can be ascribed to it. The four levels of concern are

- high concern, where without control there is a life threatening risk.

- medium concern, where there is a threat to the consumer that must be controlled.
- low concern, where there is little threat to the consumer; still is advantageous to control it.
- no concern, where there is no threat to the consumer.

The points which are not critical because of low risks need less control and monitoring. If a hazard can be controlled at more than one point, the most effective place for control should be decided.

18.7.6 Specification of Criteria for Control

The team should next identify the means by which the hazard can be controlled at each CCP. These may include level of chlorination in wash water, temperature during storage, moisture content, level of toxic materials, sensory parameters, time and temperature requirements for thermally processed foods etc. All these must be documented as clear unambiguous statements or included as specifications in operating manuals. The team will also have to set target levels and specified tolerances of the control measures at each CCP. Critical limits may also be derived from government regulations and guidelines, international codes of practices, experimental studies or any other recognised source.

18.7.7 Monitoring and Checking System

There must be a mechanism to monitor, check and measure to ensure that processing procedures at each CCP are under control. The monitoring must be able to detect any deviation from this specifications. It should be rapid enough so that corrective action can be taken in time and loss of the product is avoided or minimised. The main methods to monitor a CCP are

- visual observation.
- sensory evaluation.
- physical measurement.

- chemical testing.
- microbiological analysis.

Visual observation is basic, but gives rapid results. Sensory evaluation can be used to check the quality of incoming raw materials. Rapid chemical tests e.g., chlorine in water, are useful to monitor CCPs. Other measurements possible are time, temperature, pH and salt concentration. Microbiological test is of limited use in monitoring CCP. It can, however, be employed for testing before starting processing, and for testing finished products before release.

18.7.8 Corrective Action

If monitoring indicates that there are deviations from the critical limits or that the process is out of control, corrective actions must be taken immediately. The corrective actions must be based on the assessment of hazards, risk and severity, and on the final use of the product. A plan should have been prepared in advance so that there will be no delay in taking corrective action. The team should prepare this plan specifying the corrective action, identify the persons to implement them and disposition actions needed to be taken with the food that has been produced during the 'out of control' period.

18.7.9 Verification

Once the HACCP system has been drawn up for a product/process it must be reviewed before it is installed, and regularly reviewed while it is in operation. The appropriateness of the CCPs and control criteria can be determined and the extent and effectiveness of monitoring can be verified. The team should describe in detail the methods and procedures to be used to verify the system. Some of the methods that can be used in verification are

- reviewing the HACCP study and its records.
- random sampling and analysis (microbiological).
- detailed tests at selected CCPs.

- survey of conditions during storage, distribution, sale and use of products.
- interviewing staff.

With respect to verification of an established system, additional tests may be applied at a CCP which is of a more vigorous and searching nature, but not routinely monitored because of constraints of time. Validation of critical limits in consultation with experts, specialists and regulatory organisations, review of the HACCP system on the basis of all data collected to identify the CCPs for which monitoring system is not adequate etc. are some of the other methods of verification. If any revision is needed it must be implemented and the changes must be fully integrated into all written documents and the record keeping system.

Verification is increasingly being conducted through a risk-based audit. Quality audit by outside organisations may reveal unexpected problems. For example, examination of the end product by an outside specialist laboratory may indicate hazards that were not identified in the initial hazard analysis. Reports from market also may indicate unanticipated problems reflecting on the unsatisfactory handling of the product. These also call for immediate remedial action.

It should be remembered that the HACCP system is set up for individual product/process handled in a given way. If any change is made at any point, it may be necessary to alter the CCPs or change the methods of monitoring.

18.7.10 Documentation

Proper records should be maintained on all actions in the HACCP system in order that origin, cause and point of occurrence of the hazard can be traced. Records should be maintained on the following aspects.

- HACCP plan and supporting documentation.
- monitoring of CCPs.
- records of corrective actions.

- records of verification activities and modifications.
- nature, coding and disposition of the product.

18.7.11 Training of Personnel

Successful implementation of any programme will depend largely on the persons responsible for different activities in the programme. Individuals to whom quality assurance task is assigned should posses an in-depth knowledge of the various aspects of the discipline. This can be achieved only by education and training of the staff.

Suggested Reading

Anon (1996) *An Introduction of HACCP for Fish Processors*, 2nd edn. Asean-Canada, Fisheries Post Harvest Technology Project Phase II, Singapore.

Birch, G.G. and Parker, K.J. (1984) *Control of Food Quality and Food Analysis*. Applied Science Publishers Ltd., London.

Connell, J.J. (1990) *Control of Fish Quality*. Fishing News Books Ltd., Farnham, UK.

Skoogaard, N. (1991) *Food Laboratory News*, 7: 11.

Gerasinov, G. V. and Antonova, M.T. (1972) *Technochemical Control in the Fish Processing Industry*. Amerind Publishing Co. Pvt Ltd., New Delhi.

Gorga, C and Ronsivalli, L.J. (1968) *Quality Assurance of Seafoods*. Van Nostrand Reinhold, New York.

Huss, H.H. (1994) *Assurance of Seafood Quality*. FAO Fisheries Technical Paper No. 334, FAO, Rome.

MPEDA (1997) *A Course Manual and Guide for Implementation of HACCP Principles and Pre-Requisite Programmes with Reference to US and EU Seafood Regulations*. Marine Products Export Development Authority, Cochin.

Pearson, A.M and Duston, T.E. (eds) (1995). *HACCP Concept in meat, poultry and fish processing. Advances in Meat Research Series*, Vol. 10. Chapman Hall, London.

19
BYPRODUCTS

FISH undoubtedly is one of the most nutritious of foods available for human consumption. Fish flesh on an average contains 15-20 per cent protein. Some species of fish contain very high amounts of body oil. Few species of fish like shark, cod etc. are good sources of liver oil. However, even the best quality table fish has, even at the highest level, only 50 per cent edible flesh, the rest consisting of frame, head, and viscera, which are inedible. There is a large quantity of very small fish landed as bycatch which do not find a ready market as fresh fish. Fish processing and filleting industries turn out large quantities of fishery waste. All these are good sources of high quality protein, fat, minerals etc. The shellfish processing industry, particularly of shrimp, crab and lobster, is a major source of waste comprising head and shell. Judicious and economic utilisation of fish should have a programme of proper disposition of such fish and fishery waste by processing them into different products intended for human consumption, animal nutrition or industrially useful products.

The traditional fishery byproducts are fishmeal, fish body and liver oils, fish maw, isinglass etc. Fish protein concentrate, fish albumin, glue, gelatin, pearl essence, peptones, amino acids, protamines, fish skin leather etc. are some other byproducts generally processed out of fish and fish waste. Chitin and chitosan processed out of shrimp, crab and other crustacean waste are byproducts of high economic value. Biochemical and pharmaceutical products like bile salts, insulin, glucosamine etc. are some other fishery byproducts of great significance. A brief account of some of the important fishery byproducts is given below.

19.1 FISHMEAL

Fishmeal is a traditionally used livestock feed supplement. Fishmeal has high quality protein containing high levels of lysine, methionine and cysteine, three of the essential amino acids which the animal bodies cannot synthesise, and this makes it an unrivalled constituent of feed stuff. It is also a good source of B group vitamins like cyanocobalamine (B_{12}), choline, niacin, pantothenic acid and riboflavin. Fishmeal is rich in minerals like calcium, phosphorus, copper and iron and is also the source of some trace elements. The 'unknown growth factors', some unidentified constituents in fishmeal contributing to animal growth, is a unique feature which highlights the importance of fishmeal in animal nutrition. The biological value of fishmeal protein is very high, another factor highlighting its importance as an animal feed constituent.

19.1.1 Raw Materials

High-fat fish like anchovies, sardines, herring, menhaden etc. are traditionally used as raw materials to manufacture fishmeal. Small bycatch fish from shrimp trawling generally not marketable as fresh fish due to various reasons like very small size, bony nature, etc. also can be used. Waste from fish processing and filleting plants, cannery wastes, carcasses of fish like shark and other fish wastes are also used as raw material to manufacture fishmeal.

19.1.2 Processing

The are two methods most commonly employed for processing fishmeal, the wet rendering or wet reduction process, and the dry rendering or dry reduction process. The process selected will depend on the type of fish or fish waste used as raw material.

19.1.2.1 Wet Rendering

Wet rendering is employed almost exclusively for processing high fat fish and fish offal where simultaneous production of fishmeal and oil is envisaged. The fish or offal after grinding coarsely is cooked in live steam. The cooked mass is pressed in a screw press to expel the expressible liquid. The press cake will contain about 50-55 per cent moisture and 3-4 per cent oil. The expelled liquid, called press liquor, is passed over a shaker screen to separate the suspended solids. The wet solids thus obtained are returned to the press cake. The press cake is fluffed up to facilitate easy drying and is dried suitably, preferably in a rotary drier, to a moisture level around 8 per cent. The dried fishmeal is ground to the required particle size and is bagged for storage or disposal.

The press liquor is heated and centrifuged to separate the oil. The clarified liquor, the stick water, will still have some residual oil, about 5 per cent suspended fine solids and, dissolved minerals and vitamins. After concentration to a level of 50 per cent solid content this is often marketed as 'condensed fish solubles'. However, this concentrate can be added back to the press cake and dried along with it to produce 'whole fishmeal'.

Cooking can be done in direct or indirect steam. When cooked in direct steam, steam is introduced into the cooker through a series of pipes. In indirect cooking steam is retained in the jacket of the cooker. The conveyor screw in the cooker also can be heated by steam. Indirect cooking reduces the quantity of press liquor. Because of this, the loss of solids also is less in indirect cooking.

Cooking and pressing reduces the oil content from as high as 20 per cent in the raw fish to around 5 per cent in the press cake. Percentage of fat in the dry meat is around 6 per cent. However, meat from highly fatty fish may contain 8-10 per cent oil.

Though the feed mass for rendering is coarse ground

before cooking, it is not considered very essential step when very small fish is used. Cooking results in denaturation of proteins and breaking of cell walls so that oil and water can be easily removed under pressure.

• DRYING

There are two types of driers commonly used to dry the press cake, direct drier and indirect drier. Control of the temperature of meal particles, which is critical to nutritive value, is one of the considerations in selecting the type of drier. Direct drier is also called flame drier. This consists of a slowly rotating cylinder provided with a number of flights within. The press cake is tumbled rapidly in a stream of hot air by rotating the drier. Tumbling results in intimate contact of the material with hot air. The hot air, often at 600°C, is produced by burning any convenient fuel. However, the particles of fishmeal do not attain high temperature because of the evaporative cooling produced by rapid evaporation of water from their surface. The temperature attained by the particle is only about 80°C. The driers are generally designed to dry the wet meal to 8-10 per cent moisture in 15-20 minutes.

In direct drying the fishmeal may get contaminated with the products of combustion like soot, oxides of nitrogen, sulphur etc. or even may become charred. At some stage of drying the concentrated stick water may be added to the press cake. Proper mixing of both will often become difficult because of formation of lumps that will affect the ease of drying.

• INDIRECT DRIER

Indirect driers consist of fixed cylindrical drums provided with internal rotating scrapers or flights, which continuously stir the meal during drying. The flights are internally heated, usually by steam, and make heat transfer to the product very effective. Steam under pressure raises the temperature of the flights to 170-180°C; however, the product does not attain this temperature because of evaporative cooling. Air is drawn through the drier to remove the evaporated moisture. One cycle of drying may take 30 minutes or more which is almost double the time taken in direct driers.

Some charring of the product is likely if the material becomes stuck to the surface during drying. However, there is no appreciable difference in nutritive value of fishmeal dried in direct and indirect driers.

19.1.3 Dry Rendering

Dry rendering or dry reduction is the process employed to process fishmeal from non-oily fish and fish offal. The process is ideal for batch operation and hence can be advantageously employed even when the quantity of raw material to be processed is small. If the quantity to be processed is very small, it can be dried in the sun and pulverised. When a sufficiently large quantity is to be processed, the raw material is first coarse ground and fed into a cooker-drier. The cooker-drier is a steam-jacketed vessel equipped inside with a power-driven stirring device, usually a paddle type stirrer. The stirrer is rotated continuously, but slowly throughout the drying process. The cooker-drier may be operated under atmospheric pressure or under partial vacuum.

Dry rendered fishmeal may, sometimes, contain more oil than that is desirable in the product. In such cases the excess oil is removed by pressing the dried mass in a hydraulic press before it is ground and bagged. The oil obtained will be comparatively dark in colour and of inferior quality compared to the oil produced by the wet rendering process.

Dry rendering is a batch process with limited capacity. But the yield is higher compared to the wet process because the water-soluble materials are retained in the meal. Another advantage of dry rendering is that depending on the material handled the operating conditions can be altered.

19.1.3.1 Advantages/Disadvantages of Dry and Wet Rendering

• DRY RENDERING

Advantages

- suitable for batch operation.
- easy to manage.
- permits flexibility in the operating conditions to suit the raw materials.

Disadvantages

- oil is darker and of inferior quality and will quickly develop oxidative rancidity.
- limited capacity and slow production rate, being batch operation.
- more expensive to operate.

• WET RENDERING

Advantages

- continuous process, fast production rate.
- suitable for processing large quantities of raw material
- better quality oil.
- less expensive to operate.

Disadvantages

- lower yield.
- meal is low in water-soluble fractions unless 'fish solubles' are added back.
- rigid method of operation needing qualified operators.

19.1.4 Pulverisation

The meal coming out of the driers is coarse and differs widely in particle size. The dried mass is therefore pulverised to yield a homogeneous product. Generally the meal is milled to yield the particle size decided by its further use.

19.1.5 Curing

Some meals, particularly that processed out of fish high in oil, heats up while in storage. There are instances where the heating up of fishmeal in stacks has led to fire hazards. Oxidation of residual oil leading to polymerisation of the unsaturated fat is the phenomenon known to be involved. Fishmeal with oil content above 10 per cent and moisture content above 8-10 per cent are more prone to heating during storage; so also when unsaturation of the oil is high. It is therefore necessary to convert the fishmeal to a stable state before packing and storage. Antioxidants like butylated hydroxy toluene or ethoxyquin are widely used to prevent heating in store and the consequent loss of protein quality.

An alternate to antioxidant treatment is to 'cure' the fishmeal when the heat evolved during oxidation of fat is dissipated into the atmosphere. Bags containing fishmeal are allowed to stand singly for about four weeks and then they are stacked in such a way that the heat produced is dissipated. If stored in bulk, the bags should be turned over from time to time. At any time during the storage the temperature of the meal should not exceed 35°C.

19.1.6 Proximate Composition

Proximate composition of fishmeal will vary widely depending on the raw material used, processing conditions employed etc. The range of values is given below :

Protein	50-70%
Fat	5-10%
Ash	12-33%
Moisture	6-10%

With regard to requirements of fishmeal as an ingredient in the livestock feed, the following two grades have been recommended (Table 19.1).

Table 19.1 : Two Grades of Fishmeal as Ingredient in Livestock Feed.

Parameters	Grade 1	Grade 2
Moisture per cent, max.	10	10
Crude protein per cent, min	60	50
Ammoniacal nitrogen per cent, max	0.5	0.5
Crude fat/petroleum ether extract per cent, max	10	10
Acid insoluble ash per cent, max	3	3
Chloride (as NaCl) per cent, max	4	5

19.1.7 Uses

The principal use of fishmeal is as an ingredient in the livestock feed. Fishmeal is often supplemented at the following levels in the rations of animals and poultry :

Cattle - 907 g/day/454 kg live weight

Pig - 113-127 g/day according to weight

Sheep - 45-91 g/day/45.4 kg live weight

Poultry - Not more than 10 per cent of the total ration for hens and not more than 5 per cent for chicks

Another important use of fishmeal is in fish and shrimp feeds that has become a very important application because of the increasing aquaculture activities. The amount of fishmeal used in such feeds varies widely from 10-70 per cent. Fishmeal low in fat or stabilised with antioxidants is preferred in aquaculture feeds.

19.2 FISH PROTEIN CONCENTRATE

Fish protein concentrate (FPC) is a stable protein concentrate prepared from whole fish or other aquatic animals or parts thereof. Protein concentration is increased by removal of water, oil, bones and other materials. Traditionally dried or otherwise preserved products do not fall within this definition. Development of FPC has paved the way for converting a wide range of whole fish into protein concentrate, which has no resemblance to the original raw material, for human nutrition.

Though a number of processes including enzymatic hydrolysis have been developed for processing FPC, the most popular ones employ some organic solvent to remove lipids and water from the raw or cooked fish meat.

19.2.1 The Viobin Process

In this process, simultaneous dehydration and removal of lipids from ground whole fish is accomplished using ethylene chloride. The ground fish is first suspended in ethylene chloride. The mass is then heated to boiling using indirect steam. Ethylene chloride forms a constant boiling mixture (azeotropic mixture) with water boiling at 71°C, lower than the boiling points of ethylene chloride (83°C) and that of water (100°C). On condensation the azeotropic distillate separates into two layers. The solvent layer is pumped back into the vessel to carry off more water. As the water content approaches zero the temperature of boiling will approach that of the solvent indicating the completion of the process.

The lipids will remain in the vessel dissolved in the residual solvent. The miscella, i.e. the oil-solvent solution is filtered off and the residue is washed once or twice with fresh solvent. It is then heated in a steam-jacketed vessel under vacuum to remove as much solvent as possible. Finally, the vessel is flushed with dry steam to remove the last traces of the solvent. The dry product is then pulverised and bagged.

19.2.2 The Canadian Process

The Canadian process employs isopropyl alcohol as the solvent for extraction of lipids. The process is completed in two stages. The first stage involves grinding the fish, its suspension in water acidified with phosphoric acid (pH 5.5) and cooking for 30 minutes at 70-80°C with constant stirring. The cooked mass is filtered and washed repeatedly with hot water until the filtrate is practically odourless.

The residue from the first stage is suspended in isopropyl alcohol and refluxed for 15 minutes. The solvent is removed by filtration or centrifugation. The residue is further treated with successive batches of solvent until water and oil are reduced to the desired point. The residue after the final treatment is pressed and the press cake is dried and pulverised. Residual solvent, if any, is removed under vacuum.

19.2.3 Azeotropic Extraction Method

A process reported from India involves extraction of fat and odoriferous components from fish using an azeotropic mixture of hexane and ethyl alcohol. Minced whole fish is cooked in boiling water containing 0.5 per cent acetic acid for 30 minutes. After draining off the liquid the solid matter is pressed well and the press cake is refluxed under constant stirring for 30 minutes with an azeotropic mixture of hexane and ethyl alcohol containing 32.2 moles percent alcohol boiling at 58.68°C. After draining off the solvent mixture the solid matter is refluxed once again with fresh solvent mixture. The solvent is drained off and after distilling of the remaining solvent the extracted mass is steam-stripped under reduced pressure to drive off the remaining traces of the solvent. The mass is dried and pulverised to a fine mesh size. The final fat content in the product will be below 0.3 per cent on dry weight basis

19.2.4 Other Methods of Production of FPC

There are few other methods employed, the basic difference being in the solvents used for extraction of lipids. The British process employs ethyl alcohol whereas the Chilean process employs hexane followed by ethyl alcohol. The Moroccan process employs a solvent mixture of hexane, ethyl acetate and isopropyl alcohol.

19.2.5 Proximate Composition

Fish Protein Concentrate is a gritty, colourless, odourless and tasteless power. It is stable upto 3-4 years at room temperature without any significant change in flavour. Proximate composition of a representative sample of FPC is given below :

Moisture	8.1 per cent
Crude protein	67.7 per cent
Available lysine (as per cent of protein)	7.9 per cent
Total lipids	0.4 per cent
Calcium	5.4 per cent
Phosphorus	3.9 per cent
Ash	24.1 per cent
Pepsin digestibility	98 per cent

The high content of highly digestible protein, available lysine and minerals makes FPC a highly nutritious product.

19.2.6 Uses

Though FPC is intended for human consumption it is not relished for consumption as such. It is therefore incorporated as a protein supplement in human diet. 5-10 per cent level FPC in bread and biscuit is considered the acceptable limit. 35 g per person per day is a recommended level of use of FPC.

19.3 TEXTURED FISH PROTEIN CONCENTRATE

Fish protein concentrate, though highly nutritious, has the limitation of lacking in palatability. This is particularly because of its low rehydration ability making it difficult to process with other foods. To overcome this various attempts have been made to produce an FPC with improved rehydration ability, i.e., FPC with improved functional properties called textured fish protein concentrate. The product is also known as marine beef. In processing FPC the whole fish after grinding is treated with organic solvents to remove fat and water and then dried and pulverised. In the process for textured protein concentrate only the meat is used which is kneaded with salt to make a paste which is then treated with ethyl alcohol at low temperature. The dry product will have good hydrophilic property.

19.3.1 Processing

The meat separated from the fish is first washed. Meat from low fat fish is given a light wash whereas that of fatty fish is washed thoroughly with a 0.4 per cent aqueous solution of sodium bicarbonate. The excess water from the washed meat is removed either by centrifuging or under pressure. The meat is then mixed with 0.5-1 per cent sodium chloride by weight of the meat and the pH is adjusted to 7.4-7.8 using sodium bicarbonate. After adding further 1-2 per cent more of sodium chloride the meat is kneaded to a viscous paste. This is mixed with 3 times its volume of ethyl alcohol at 5-10°C and kept stirred for 15 minutes. The coagulated meat is extruded in the form of long strands. Alcohol is removed by centrifuging and the residue is again passed through the extruder and mixed with 3 times its volume of cool ethyl alcohol for 15 minutes. After removing the alcohol by centrifuging the residue is dried in hot air at 30-45°C to a moisture content less than 10 per cent. Products with different texture and rehydration capacity can be produced depending upon the number and length of treatment with alcohol or by changing the temperature of the alcohol.

19.4 FISH OILS

Fish oils are of two types, the liver oil and the body oil. Fish liver oils were used for therapeutic purposes in the treatment of vitamin A and D deficiencies. The most important sources of liver oils are cod, haddock and shark. Liver oils of halibut and tuna also are rich sources of vitamins A and D.

The body oil of fish is more important as an industrial

product besides its limited use in human nutrition. Body oils have recently won much attention because of the content of polyunsaturated fatty acids (PUFA), particularly n-3 PUFA used in the control of heart ailments in humans.

19.4.1 Liver Oils

Therapeutic value of fish liver oils is known since very early days. They are the most important sources of natural vitamins A and D. However, with the advent of synthetic vitamin A at economic price, liver oils lost their importance as source of vitamin A. Still they are considered good source of vitamin D.

19.4.1.1 Classification of Livers

Weight of the liver, fat content and vitamin content depend on the species of fish, their age, nutritional status and stage of spawning. In cod and haddock, the liver accounts for 4 –9 per cent by weight of the whole fish, and the liver contains as high as 45-70 per cent oil. Liver accounts for as much as 10-25 per cent of the body weight in some species of sharks which have large fatty liver having very high oil content in the range 60-75 per cent. In halibut, tuna and whale the liver amounts to 1 per cent of the body weight and contains 4-25 per cent oil having high vitamin A and D potency.

Depending on oil and vitamin contents fish liver is classified into three groups viz.,

High oil - low vitamin A potency
Low oil - high vitamin A potency
High oil - high vitamin A potency

19.4.1.2 Preservation of Liver

The golden rule to produce high quality light coloured oil with good odour, flavour and, containing minimum free fatty acids is that the liver be removed from the fish and processed as quickly as possible after catch. However, due to various reasons, this may often be not possible. Therefore preservation of liver for varying periods may become necessary. Fish liver is very susceptible to bacterial and lipolytic degradation. The method of preservation chosen shall take into consideration these aspects as well as the length of preservation needed. The liver may be separated from the fish and preserved by one of the following methods:

- store in crushed ice in cold room. This will preserve the liver for a couple of days.
- store in frozen storage maintained at -18°C or below. This can ensure safe storage for several months.

Other methods suggested for temporary preservation of fish liver include the following :

- chopping the liver into small pieces, mixing with 10 per cent by weight of common salt and storing in air-free air tight container.
- grinding the liver and preserving with 0.25 per cent of its weight of formalin.
- grinding the liver and keeping in aqueous solution of 2 per cent sodium carbonate containing some germicide like formaldehyde, phenol, resorcinol or alcohol.

19.4.1.3 Extraction of Oil

As fish livers differ widely in their oil content and vitamin potency the process employed also may differ to suit the particular type of liver. Extreme care is needed in processing liver with low oil content of high vitamin potency where the oil and the vitamin are held closely connected to the proteinaceous tissues of the liver.

• LIVER HAVING HIGH OIL - LOW VITAMIN A POTENCY

Liver of fish like cod, haddock and similar species contains 50-70 per cent oil by weight. The vitamin A potency ranges from 500-20,000 U.S.P. units per gram oil. The large volume of oil acts as a solvent for vitamin A thereby facilitating its extraction. Direct or indirect steaming processes can be employed for extraction of oil from livers of this class.

(a) Direct steaming method

Direct steaming is the simplest method for extraction of oil from livers of the above class. The process involves introducing steam directly into the cooker containing the liver and raising its temperature to 80-86°C. Cooking results in thermal rupture of liver cells releasing the oil. Heating is continued until the liver disintegrates completely and the oil floats on the top. The oil is skimmed off and kept in a settling tank or centrifuged to remove suspended solids and moisture.

(b) Indirect steaming

The liver is kept in a steam-jacketed kettle and heated indirectly by steam admitted into the jacket. The temperature is maintained at 70-75°C. The material will be kept mechanically stirred throughout to facilitate disintegration of the liver and release of oil. The oil liberated is separated as in the previous case.

• LIVER HAVING LOW OIL-HIGH VITAMIN A POTENCY

Liver of halibut, tuna, lingcod and certain species of shark contain 4-28 per cent oil by weight and vitamin A potency in the range 25,000-600,000 U.S.P. units per gram oil. As the oil is held more closely by the proteinaceous liver tissues, simple steaming techniques will not release the oil without degrading vitamin A. Therefore, a process of digestion or solubilisation of protein is employed to release the enclosed oil and vitamin A. The processes employed are (a) alkali digestion, (b) enzyme and alkali digestion, (c) acid digestion and d) solvent extraction.

(a) Alkali digestion

Liquefaction of liver using a mild alkali is the simplest

method applicable for releasing oil high in vitamin A potency from the livers low in oil content. The liver is first ground and mixed with 1-2 per cent by weight of sodium hydroxide or 2-5 per cent by weight of sodium bicarbonate. The mixture is heated with live steam maintaining the temperature at 82-87°C. The mixture is kept stirred while cooking is continued. The digested liquor is centrifuged to separate the oil.

In the alkali digestion process it is very significant to maintain the correct level of alkali sufficient for digestion, but not above, to prevent saponification of oil and absorption of vitamin A in the soap fraction.

(b) Enzyme and Alkali Digestion

This is a process where alkali digestion is preceded with an enzymic hydrolysis. The liver is ground and mixed with an equal volume of water. The pH of the slurry is adjusted to 1.2-1.5 using hydrochloric acid. An aqueous suspension of 0.05 per cent papain (based on the weight of liver) is added and the mixture is maintained at 43 – 48°C for 35-48 hours. The mixture is stirred constantly during hydrolysis. At the end of hydrolysis the pH of the mass is adjusted to 9 with saturated sodium carbonate solution and is further subjected to alkali digestion for another one hour at 80°C. The floating oil is recovered by centrifuging.

(c) Acid Digestion

The acid digestion method involves grinding the liver, adjusting its pH to 1.5 with an acid and cooking under constant stirring. The oil is separated from the digested mass by centrifuging.

(d) Solvent Extraction

The liver has to be disintegrated and desiccated for efficient recovery of oil in the solvent extraction process. The liver is, therefore, first ground and as much water as possible is removed by steaming the ground mass at 70-75°C for an appropriate length of time, generally 30-45 minutes, under stirring followed by draining off the released water while the mass is still hot. The residue is covered with a layer of paraffin or carbon dioxide gas and is rapidly cooled to around 20°C. The oil is extracted with one of the solvents like acetone, ethyl ether, dioxane, benzene etc. The solvent with the extracted oil is filtered and vacuum distilled to recover the oil.

• LIVER HAVING HIGH OIL-HIGH VITAMIN POTENCY

Livers coming under this class are grouped into two, viz. livers containing 30-75 per cent oil with vitamin A potency of 0-340,000 U.S.P. units per g and livers containing 45–75 per cent oil with vitamin A potency of 20,000–200,000 U.S.P. units per gram. Because of these wide variations, the process employed will differ depending upon the characteristics of the liver.

19.4.2 Body Oil

Body oil is extracted from whole fish or fish offal by the wet reduction process (section 19.1.2.1). Wet rendering is particularly suitable for high fat fish like oil sardine, pilchard, herring and the like. The process can also be advantageously employed when fish low in oil content like trawler bycatch is used where recovery of oil also is intended. Oil extraction is never attempted when the oil content in the fish is very low. However, if oil is recovered from fishmeal processed out of such fish the ideal method is to recover it from the meal using a hydraulic press or by solvent extraction.

19.4.2.1 Wet Rendering

The stick water (section 19.1.2.1) is reheated, if necessary, and then centrifuged and separated into fish body oil and fish soluble. Oil can be separated from stick water by using settling tanks. A series of tanks, generally four, are used where separation of oil and further purification with hot water are achieved in successive stages.

19.4.2.2 Dry Rendering

If there is enough oil in the meal justifying its recovery, the meal can be pressed in hydraulic press. However, the oil is likely to be inferior to the oil processed by wet rendering process, particularly in the colour that may get darkened by reaction with the heated metal surface of the cooker drier.

19.4.2.3 Solvent Extraction

Solvent extraction is an expensive process and hence it is not popular in extraction of fish body oil. In addition, the oil extracted by this process is considered inferior to the oil produced by wet rendering process. If needed, the same process as applied for extraction of liver oil can be employed for body oils also.

19.4.3 Refining

Crude fish body oil may contain several undesirable materials in varying amounts. These include suspended matter, free fatty acids, naturally occurring oil soluble colouring matter, volatile odoriferous and flavour bearing compounds and saturated glycerides. These have to be removed by some refining process to increase the utility of oil in industrial application. Different methods of refining are employed which may serve one or more functions at the same time. The following methods are employed for refining fish body oils.

19.4.3.1 Winterisation

A method of refining crude oil practised since ancient times and still popular is winterisation, also called 'cold clearing'. When saturated or other high melting glycerides are present in the oil, they crystallise at refrigerated temperature. High melting glycerides usually contain substantial quantities of stearic acid and hence these glycerides separating on cooling are called 'stearines'. Winterization involves holding the oil at refrigerated temperatures, generally at 5°C, until crystallisation is well advanced. The saturated triglycerides can be separated by filtration in a chilled room. The clear oil obtained will have improved drying properties. The process is simple and easy to practice.

19.4.3.2 Gravity Settling

The oil is maintained in a fully liquid state for several days in settling tanks. The coarse suspended matter present in the oil settles at the bottom of the tank by gravity. Clear oil is drawn off from the top.

More often, the oil contains colloidal and fine suspended matter. To clear the oil from such matter it is first blown with moist steam at 100°C for some time and then allowed to cool. Addition of common salt breaks the emulsion formed. After settling, the condensed water and oil water emulsions are centrifuged to separate the oil. The process yields cleaner oil.

19.4.3.3 Alkali Refining

A very effective and widely employed method for complete removal of free fatty acids and various colour producing materials from oil is alkali refining. A dilute solution of sodium hydroxide is added to heated oil and stirred vigorously. A slight excess of alkali is always added to ensure complete neutralisation of free fatty acids. Instead of sodium hydroxide, concentrated solution of sodium carbonate also can be used. After short period the mixture is allowed to settle. The settlings called 'foots' are collected and sold as 'soap stock'.

In a continuous process, appropriate quantities of caustic soda solution and oil are continuously blended at 20-30°C and is then heated to 55-70°C in a heat exchanger to break the emulsion. The mass is then centrifuged to separate the 'foots'. The oil is washed free of alkali and is again centrifuged to separate the clear oil.

19.4.3.4 *Bleaching*

Some fish body oils exhibit a natural greenish or reddish colour due to the presence of caroteniod or chlorophyll pigments. Since most of the fish body oils are highly unsaturated, use of chemical bleaching agents is considered not desirable. Adsorptive bleaching techniques like treatment with Fuller's earth, activated charcoal etc. are popularly employed in such cases. The oil is heated with the adsorbent to 120-140°C for about an hour. The oil is then separated from the bleaching material by filtration.

19.4.3.5 *Deodourisation*

The characteristic odour of fish body oil is considered to be due to the presence of oxidation products of highly unsaturated fatty acids, free or in the triglyceride form. This gain support from the fact that when oil is hydrogenated, such odours are lost.

Deodourisation is essential for advantageous use of fish body oils for edible purposes, particularly in improving the flavour of hydrogenated fish oils. Hydrogenation is very effective as a process for deodourisation. Another method employed is placing heated oil in a vacuumed tower and allowing it to cascade over steam in countercurrent direction. This process removes volatile odoriferous substances like aldehydes and ketones and also destroys peroxides and carotenoid pigments.

19.4.4 Deteriorative Changes in Fish Body Oil

Because of the highly reactive nature contributed by the presence of high proportion of unsaturated fatty acids, fish body oils undergo several deteriorative changes, post-mortem in the fish or after extraction and storage. The important among them are development of free fatty acids, oxidative rancidity and flavour reversion.

19.4.4.1 *Free Fatty Acids (FFA)*

Free fatty acids (FFA) are produced by the action of lipolytic enzymes present in fish tissues, particularly in the entrails. These enzymes become very active after the death of the fish. Contamination of fish oil with micro-organisms, presence of water even in very small quantities in oil etc. will accelerate the lipolytic activity. Lipolytic activity continues at ordinary temperatures of storage, but can be prevented by storing at very low temperatures, say at –10°C.

19.4.4.2 *Oxidative Rancidity*

Rancidity develops due to the action of atmospheric oxygen or lipoxidase enzymes. Atmospheric oxygen reacts with unsaturated oils even at ordinary temperature and pressure. At elevated temperatures and in the presence of a catalyst, even saturated fatty acids react with oxygen most readily.

Lipoxidases, the naturally occurring enzymes in fish tissues, also produce oxidation products of oils. Lipoxidases can also be produced by bacteria in the presence of as little as 0.3 per cent moisture. Lipoxidase enzymes are inactivated by heating at 80-100°C and hence is not considered a serious problem in oil extracted using heat.

19.4.4.3 *Flavour Reversion*

Deodourised fish oils, on storage, develop a fishy odour and flavour similar, or not, to the original fishy flavour. The phenomenon is called 'flavour reversion'. It is caused primarily due to the oxidation of the deodourised oil. Chemical combination of nitrogenous compounds with highly unsaturated triglycerides in the presence of peroxides also is considered contributing to the phenomenon of flavour reversion.

19.4.5 Control of Deterioration in Fish Oils

Most important method of controlling deterioration of

fish oils is the use of antioxidants, either synthetic or naturally occurring. Hydroquinone, pyrogallol, catechol, para amino benzoic acid, butylated hydroxy anisole etc. belong to the former group. Tocopherols, phosphatides, gallic acid, gum guaiac, nordihydroguiaretic acid, ascorbic acid etc. are the antioxidants under the latter category. Combinations of antioxidants having synergetic effects can be more advantageously employed to prevent oxidation.

Other methods to control deteriorative changes in fish oils include stabilisation by means of a very slight halogenation and storing under an inert gas like nitrogen.

19.4.6 Uses of Fish Body Oils

Highly unsaturated fish body oils can be used as drying oils in paints and varnishes. Many oils are used for human consumption as cooking oil after hydrogenation, and also as a medium in fish canning. It also finds use in margarine. They are used as carriers of fat-soluble vitamins A and D. They also find use in the manufacture of linoleum, detergents, artificial rubber, lubricants, printing inks, soaps etc.

Body oil of sardine and similar species are good sources of polyunsaturated fatty acids (PUFA), particularly n-3 PUFA which are known to have anticholesterolemic effect and hence is largely used in control of heart ailments in humans. n-3 PUFA concentrates are now available in capsule form.

19.5 GELATIN

Gelatin is a protein that lacks in an essential amino acid tryptophane, and hence cannot be considered as a sole source of protein in animal or human nutrition. But it is a relatively high source of lysine and methionine, which are deficient in cereal proteins. However, gelatin finds extensive use in food as also in the formulation of some industrial products. Gelatin can be extracted from the skin and bones of fish.

19.5.1 Extraction

Raw fish skin is washed in running water for 3 to 4 hours. After draining off the water, it is soaked in dilute sodium hydroxide, not more than 0.5 per cent concentration, for 6 to 8 hours. The low alkalinity is maintained in order to avoid too much swelling of the stock. It is then washed in running water for 3-4 hours and macerated three times in fresh solutions of weak sulphurous acid, followed by thoroughly washing in running water. Extraction of gelatin is carried out by adding two parts of water to each part of pre-treated skin and heating to 70-80°C for two consecutive 30-minute periods.

Gelling properties of fish gelatin is not as good as that from land animals. However, fish gelatin can be advantageously processed and used in coastal countries which have very small animal industry.

19.5.2 Uses

Gelatin is used in the food industry as a gelling, stabilising, emulsifying, dispersing or thickening agent. It is used in photo engraving and chemical etching of metal parts. It has applications in the optical industry in the formulation of coating for light sensitive materials like blue print papers.

19.6 FISH GLUE

Fish gelatin and fish glue are more or less the same except that the former is of a high grade, light in colour and yields solutions that are reasonably clear and sweet. Fish glue can be made from fish skin and head. If needed, fish skin can be preserved by salting, if for short period, or by drying, for longer periods, before processing into glue. However, fish head should be processed fresh.

19.6.1 Processing

19.6.1.1 Glue from Fish Skin

Skin, whether fresh or salted, is washed and soaked in

fresh water for periods extending from 1-2 hours for fresh skin to as much as 18 hours for salted skin. The washed skin is kept immersed in a dilute (0.2 per cent) solution of caustic soda or saturated lime to open the fibre bundles and remove cementing material. It is then neutralised with hydrochloric acid and washed again in cold running water. The skin becomes swollen by now. The treated skin is transferred to a steam-jacketed double-bottomed kettle, is covered with an equal weight of water and is heated with steam. Small quantities of acetic acid, 2-4 litre per tonne of fish also may be added to the mixture to hasten the hydrolysis of the stock into glue and to act as a catalyst. Cooking is continued for about 8 hours and the glue liquid is drawn off from the bottom of the cooker. A second run may be made in a similar manner.

The liquid glue is concentrated in open heated pans at atmospheric pressure until it reaches the required viscosity or the solid content is 50-55 per cent and is then cooled. Small quantities of volatile essential oils may be added to preserve the glue and to mask the fishy odour.

19.6.1.2 Glue from Fish Head

Fish head should be processed fresh. Some bleaching agent like sulfurous acid or sodium bisulphite is added while processing fish head. Whereas addition of some glacial acetic acid is considered desirable during cooking of skin, addition of glacial acetic acid in substantial quantities, say 4.5-9 litres per tonne of head, is an essential requirement in processing fish head. Acetic acid is particularly useful to soften the head bones. Larger amounts of preservatives and essential oils are needed than for preservation of the glue from skin.

19.6.2 Uses

Fish glue is used in furniture, box making, sizing, special cements, production of half tone plates for photoengraving, book binding, small repair work etc.

19.7 FISH MAWS AND ISINGLASS

The air bladder, also called sound or swim bladder, of fish consists of several membranous layers rich in collagen. Located in the abdominal cavity below the vertebral column, air bladder helps the fish in regulating its specific gravity thus enabling it to survive at different levels in water. Cleaned and dried air bladder is called fish maw. Fish maw on further refining yields isinglass, which is an excellent raw material for producing good grade gelatin or glue.

19.7.1 Processing

The air bladder, after washing well in water and scrapping off the outer layer, is split open longitudinally and washed well further. It is then dried in sun to a final moisture level of around 15 per cent by hanging or placing in trays. The dried product is fish maw.

The dried air bladder is immersed in water until it becomes soft. Soaking for several hours may ordinarily be needed. It is then rolled, often after cutting into small pieces, between water cooled iron rollers to convert it into thin strips or sheets 3-6 mm thick. They are then further compressed by ribbon rollers into ribbons about 0.4 mm thick. These ribbons are air dried and rolled into coils. This is isinglass

19.7.2 Uses

Isinglass swells uniformly in water producing a fibrous structure not present in other gelatins. By virtue of this property, isinglass is used as a clarifying agent for beer, cider, wine, vinegar etc. The suspended impurities get enmeshed in the fibrous structure and settles. The clear liquid can be separated by decantation. Isinglass can be used also as an adhesive base. Isinglass dissolved in acetic acid forms a strong cement base useful in glass or pottery. It can be used as a sizing agent in textiles and as an ingredient in Indian ink.

19.8 PEARL ESSENCE

Pearl essence is a suspension of crystalline guanine in water or an organic solvent. Guanine is an iridescent material found in the epidermal layers and scales of most pelagic species of fish like oil sardine, mackerel, herring etc. When guanine particles are deposited on the inside surface of solid beads, an optical effect similar to that of real pearl is obtained. In its crystalline form guanine reflects and refracts light and thus acts as a camouflage to fish.

19.8.1 Isolation

Guanine deposit on the fish scales is more readily recoverable compared to that on epidermis. Freshly removed scales are collected and washed to remove adhering foreign matter. The epidermis also, if obtainable, is washed along with the scales. Scales can be preserved, if needed, in 10-15 per cent brine. The brine is later drained off and the scales are squeezed in muslin cloth bags and compressed. The compressed mass can be stored at 0°C for several weeks. It should be ensured that at no time the scales are allowed to dry. Pearl essence can be prepared as an aqueous or non-aqueous suspension.

19.8.1.1 Aqueous Suspension

Washed scales are agitated with minimum quantity of water containing little ammonia in an agitator similar to a domestic washing machine. The mixture is then passed through a strainer to remove the scales. The pearly substance present as a suspension in the liquor is purified by settling in a cool atmosphere. Guanine settles and the supernatant is decanted and replaced with fresh ammoniacal water. The process is repeated several times until the guanine crystals are fairly well purified. 0.3 per cent salicylic acid is used as preservative. Adhesives like fish glue or isinglass also may be dispersed in the aqueous suspension.

19.8.1.2 Non-aqueous Suspension

Non-aqueous suspension of guanine is made by suspending it in organic solvents like acetone, amyl acetate, chloroform and carbon tetrachloride, or in acetic acid or acetic anhydride. It is also presented in the form of a thick paste of crystals, suspended in a viscous liquor of cellulose in amyl acetate.

19.8.2 Uses

The most important use of pearl essence is in the manufacture of artificial pearls. It is used as a spray or dip for several items to impart an iridescent sheen reminiscent of pearls. It is used on such diverse articles as shoe, pencil, fishing rod, spectacle frame, walking stick, ash tray, vanity bag, book cover and even in finishes for textiles.

19.9 INSULIN

Insulin is a hormone used for correcting the condition called diabetes mellitus in humans. This is produced by the islets or caps located on the epithelial tissues on the pancreas of cattle. These islets are called 'islets of Langerhans' so named in honor of the person first to describe them. But in many fish, insulin is not secreted in the pancreas, but is associated with gall bladder or bile duct. This insulin can be more easily isolated in high purity in relatively high concentrations than that from cattle. Fish insulin is more stable as it is not subjected to decomposition by protein splitting enzymes of pancreas.

19.9.1 Isolation

Fish like cod, pollock, halibut etc are good sources of insulin. Soon after capture these fishes are gutted and the entrails are collected. The caps or islets containing insulin located on the gall bladder are identified and are clipped off with a pair of scissors. The tissues are preserved in 95 per cent alcohol containing 0.3 per cent hydrochloric acid. A more desirable alternative is to freeze the tissues using solid carbon dioxide.

From the preserved tissues alcohol is filtered off and the residue is squeezed dry. It is ground well and re-extracted in the supernatant alcohol which is adjusted to a 75 per cent alcohol concentration in the mixture of tissues and fluid. After complete maceration the mixture is kept at room temperature for about 1.5 hours and the alcohol is again drained off. The extraction is repeated thrice and the combined extracts is filtered and distilled under vacuum. The residual oil if any found in the aqueous insulin solution is removed by extracting with ether.

19.10 FISH ALBUMIN

Fish albumin is a product similar to egg albumin in physical and chemical properties. It can be processed out of proteinaceous residue from fish scrap or fish waste. Two grades of fish albumin are produced, the technical grade and, the food and pharmaceutical grade.

19.10.1 Technical Grade Albumin

The fish waste is ground well, suspended in water containing 0.5 per cent acetic acid and heated for one hour at 70-80°C with stirring. This brings about partial hydrolysis and extraction mainly of connective tissues. The protein that is not extracted is then washed in cold water and pressed sufficiently to remove about 40 per cent water in the press cake. Any oil, if present, in the cake is extracted with ethyl alcohol or trichloroethylene. The defatted mass is dried under vacuum for 2-3 hours at 50°C. The dried material is technical grade albumin.

19.10.2 Food and Pharmaceutical Grade Albumin

To produce food and pharmaceutical grade albumin, the technical grade albumin is finely ground and is digested in a solution of caustic soda. For every 100 kg dry albumin

6-8 kg caustic soda and 500 litre water are used. The digestion is carried out initially at 30°C for one hour followed by another hour at 80-90°C. After digestion, the mass is neutralised with acetic or lactic acid. The neutralised product is spray dried at 80°C. Food grade albumin is 100 per cent digestible and contains mostly polypeptides with little or no free amino acids. Neutralisation with acetic acid yields a product, which is not hygroscopic. The product neutralised with acetic acid, however, will have a better flavour.

19.10.3 Uses

Fish albumin is widely used in food and pharmaceutical products as whipping, suspending or stabilising agent. Technical grade albumin performs similar functions in industrial products.

Food grade albumin is an additive in ice cream, soup powder, puddings, confectionery, bakery products, mayonnaise, custard powder etc. Technical grade albumin finds application in paints, varnishes, textiles, paper, synthetic resins, leather, lacquer, foam rubber, cosmetics, soap etc.

19.11 PEPTONE

Peptone is an intermediate product of hydrolytic break down of protein, the products in the different stages being proteases, peptones, polypeptides and finally amino acids. Peptones are soluble in water, and are not heat coagulated. Because of this property peptones are extensively used in bacteriological culture media. For such uses peptones are prepared by careful partial hydrolysis of complete proteins, i.e. proteins containing all amino acids, usually meat or casein. Fish proteins contain all amino acids in a balanced form and hence form an ideal material for production of peptones for bacteriological culture media.

19.11.1 Process

Fish proteins can be hydrolysed by any one of the four popular methods employing peptic enzyme, tryptic enzyme, acid or alkali. The first step in the process is common to all four methods and consists in grinding the fish flesh and diluting with an equal weight of water and making into a slurry.

19.11.1.1 Peptic Enzyme Hydrolysis

Fish meat slurry in water is heated at 100°C for 5-10 minutes to inactivate vegetative bacterial cells. It is then cooled to 35°C. The slurry is acidified to pH 2.0 with dilute hydrochloric acid. Fish intestinal mucosa that contains peptic enzymes or enzyme pepsin is then added and the mixture is incubated at 35°C for about two hours for partial hydrolysis to the peptone stage. The mixture is centrifuged and the hydrolysate is heated to 80°C to inactivate the enzyme. After cooling, the hydrolysed mass is neutralised using sodium hydroxide. It is finally filtered and concentrated to a level of 30 per cent solids at a temperature not exceeding 100°C and then spray dried.

19.11.1.2 Tryptic Enzyme Hydrolysis

The slurry is heated to 100°C and cooled as in peptic enzyme hydrolysis. pH is adjusted to 8.5 with dilute sodium hydroxide solution. Fish pyloric caeca that contains tryptic enzymes is added and the mixture is incubated at 35°C for several days or weeks until the hydrolysis proceeds to the peptone stage. After cooling the slurry is centrifuged, the hydrolysate is filtered, concentrated and spray dried.

19.11.1.3 Acid Hydrolysis

pH of the slurry is adjusted in the range 0.5-1.5 with hydrochloric or sulphuric acid and is hydrolysed through the peptone stage under steam pressure at 121°C for about 5 hours. The hydrolysed mass is filtered and the filtrate is neutralised using ion exchange resin to remove the acid. The solution is concentrated at a temperature not exceeding 100°C and then dried under partial vacuum.

19.11.1.4 *Alkali Hydrolysis*

Dilute solution of sodium hydroxide replaces acid in alkali hydrolysis process. After filtration the alkali is removed by ion exchange resin. It is then concentrated and dried under partial vacuum.

Among all the methods of hydrolysis, enzymic hydrolysis is the best process inasmuch as none of the amino acids is destroyed. Acid and alkali hydrolysis partially or wholly destroys some amino acids. However, the fish peptones produced by all the above methods resemble commercial peptones in colour, odour and appearance.

19.12 AMINO ACIDS

Amino acids are the building blocks of proteins. Fish proteins have very high nutritive value as they contain all the essential amino acids as also non-essential ones in a highly balanced pattern and are readily digestible and assimilable. Fish protein can therefore serve as a good raw material for the preparation of amino acids.

19.12.1 Preparation

Amino acids can be prepared from fish protein by hydrolysis with alkalis, acids or enzymes as in the case of peptones. Hydrolysis should proceed to the stage of amino acids. As in the case of peptones, enzymic hydrolysis is considered a better method as other methods are known to partially or wholly destroy some of the essential amino acids in fish proteins.

19.12.2 Uses

Amino acids find use in the medical field as well as in food. Amino acids and their salts are administered to patients who suffer from gastrointestinal disorders or who are underfed for long periods. However, amino acids have a disagreeable taste when taken orally. Therefore, they are better administered by intravenous injections.

19.13 FISH SILAGE

Fish silage is liquefied fish protein. Preservation of surplus fish and fish offal as silage for use in animal feeding is an alternative to processing fishmeal. Ensiling does not require heavy machinery, equipment or high investment. The process is very simple and is suitable for operation with any quantity of fish. It is quite suitable for operation even in remote fishing villages where the infrastructure for any type of processing is limited.

19.13.1 Processing

Fish silage can be made from whole fish or parts of fish by treating it with mineral acid (sulphuric acid) or organic acid (formic acid). Ensiling can also be achieved by lactic acid produced by fermentation of sugar using lactic acid bacteria or starter culture. As the source of sugar, molasses is added to the fish or fish offal. If the fish has more than 20 per cent oil it must be removed to reduce rancidity and the flavour imparted to the flesh of the animal fed on silage. In all the processes the fish is partially digested and preserved by the acidity of the medium.

19.13.1.1 Acid Fermentation

Fish is ground to particle size not more than 10 mm in diameter. The ground material is mixed well with 3.5 per cent by weight of formic acid of 85 per cent concentration using a mechanical mixer. The quantity of acid used must be sufficient to bring down the pH of the mixture to 4. Mixing must be thorough enough not to leave any pocket of material uncovered by the acid. Fermentation is faster at slightly high temperature, but above 40°C the enzymes will become inactive. Periodic agitation will assist the process of liquefaction of protein that may take 1-2 weeks depending upon the freshness and oil content of fish, and also the temperature. Fresh fish will liquefy more quickly than stale fish, fatty fish more quickly than lean fish and bony fish more quickly than cartilagenous fish.

Acid fermentation can be brought about by using mineral acids like hydrochloric or sulphuric acid. In such cases the pH is reduced to below 2 by adding acids. However, the acid level requires neutralisation before feeding which is done by adding calcium carbonate at about 1-5 kg per 100 kg silage. Formic acid fermentation has the advantage that no neutralisation is needed before feeding the animals.

19.13.1.2 Fermentation by Lactic Acid Produced by Bacterial Action

The ground fish is made into a slurry by mixing with 10 per cent by weight of molasses and 30 per cent by weight of water. It is cooked for 10 min and then cooled. 18-22 hours old culture of *Lactobacillus plantarum* is introduced into the slurry, mixed well and allowed to ferment for 72 hours. The shelf life of the product is very high, even few years.

19.13.2 Separation of oil

The fish silage has the same composition of the fish from which it is made. Oil content may be high if oily fish is used for fermentation. Therefore some de-oiling may be needed to adjust the oil content to the desirable limits in the finished feed incorporated with ensilage. De-oiling is done immediately after liquefaction. Oil is often removed by centrifuging the liquid mass.

19.13.3 Uses

Fish silage is used as cattle feed. Either the whole mass or the decanted liquid portion can be used. When solid feed is desired, the silage is mixed with rice bran or other feed ingredients.

19.14 FISH SKIN LEATHER

Leather is the skin or hide of animals chemically processed to protect against deterioration by microbes.

Leather is, by and large, processed out of skin of land animals. However, the skin of certain species of fish and other aquatic animals like shark, salmon, cod, porpoise, dolphins etc. also can be converted into good quality leather. Shark skin leather is a byproduct of shark fishery where shark liver is still used as a source of natural vitamin A.

19.14.1 Processing

Skin is carefully removed from shark or other suitable species of fish, either on board the vessel or immediately after landing. Skin should be free from defects like scars, butcher cuts, wrinkles or excessive variation in thickness. Adhering meat is scraped off and the skin is cured with salt by spreading it over the flesh side. Alternately, skin can be preserved by drying for processing later.

If the skin is dried/salted, it is first washed and soaked in water. This step restores the original state of texture and consistency to the skin. It is then limed in a saturated solution of calcium hydroxide containing some sodium sulfide. Liming softens the epidermis facilitating its easy removal. The skin is subjected to several fresh liming operations followed by washing in warm water. The washed skin is next subjected to bating, a process of hydrolysing the elastin fibres in the skin using proteolytic enzyme, particularly trypsin. If not hydrolysed, elastin fibres will interfere with the proper swelling of the skin. After bating, the skin is tanned using one of the established process, vegetable or chrome tannage. When vegetable tannage is employed, shagreen should be removed before tanning. Fat liquoring, dyeing, drying and finishing operations follow to give the leather its final shape, texture and consistency.

19.14.2 Uses

Fish skin leather is used in the manufacture of such fancy items as wallets, chappals, ladies' handbags etc.

19.15 SHARK FIN RAYS

Another valuable byproduct from shark is the shark fin. Shark fins are valued for their rays, which is an essential ingredient of certain exotic soups. The fins are graded according to size and the content of the rays. On this basis fins are broadly classified into two varieties, generally known as black and white. The yield of rays from the black variety is only around half of that from the white variety. Within the same variety wide variations occur between yields of rays from the fins from different body parts. Dorsal and ventral fins yield almost double the quantity of rays as obtained from the pectoral fins. Dorsal, ventral pectoral and caudal fins are used for extraction of rays.

19.15.1 Separation of Fins and Drying

Fins are cut off at their base from sharks 125 cm or more long. Any adhering flesh is trimmed off and washed well with water. The fins are then dusted with common salt at the rate of one kg for every 10 kg fins, the cut portion receiving a liberal sprinkling with salt. Little lime is also dusted on the cut portion. The fins are then set aside for 24 hours. After gently washing in clean water to remove solid salt and excess lime the fins are dried in sun to a moisture level of 10 per cent.

19.15.2 Processing for Rays

The dried fins are soaked overnight in 10 per cent solution of glacial acetic acid in water. The shagreen is scraped off using a knife. After washing off the adhering scrap residue with water, the fins are further soaked in the same acid till they become soft. Soaking may extend to 4-6 days depending on the dryness of the fin as well as their length of storage in dry condition. The skin and the softened muscle are then scraped off and the rays are separated individually from the flesh while constantly being washed under water. Alternately, if the rays are required to be retained in the form of a cluster, some flesh enough to hold the rays together is retained at the base. The rays are then washed until free from acid and dried in sun or in an air drier at 55-60°C to a final moisture level of 10 per cent.

19.16 FISH CALCIUM

Calcium powder processed from the backbone of tuna can be used to combat calcium deficiency in the diet, particularly of children. Calcium deficiency can lead to bone failure and spine curvature in children.

19.16.1 Processing

The method of production of calcium involves mainly removing the gelatin from the crushed bones and pulverising the remaining portion. A process recommended for processing calcium powder from the backbone of skipjack tuna involves the following steps. The bone frame is crushed and washed in clean water a number of times. A 10 per cent solution of calcium carbonate is added to the residue and is left for an hour. After draining the solution, washing and treatment with calcium carbonate is repeated a number of times. Finally the bone residue is washed, dried and pulverised to the required mesh size.

19.17 SHARK CARTILAGE

Shark cartilage assumes importance because of the presence of chondriotin sulphate which is a mucopoly-saccharide. Chondriotin sulphate has therapeutic uses and is effective in reducing cancer related tumours and inflammation, and pain associated with arthritis, psoriasis and enteritis. Oral intake of shark cartilage is reported to be effective in the above cases.

The bones separated from the shark is cleaned for removing the adhering meat, blood stain etc. After washing well the bones are preserved by drying at a temperature not exceeding 70°C to a moisture level below 6 per cent.

19.18 FERTILIZER FROM FISH WASTE

Fish is a good source of essential nutrients like nitrogen, calcium, phosphorus and other elements essential for plant growth. Fish or fish waste, being organic in nature, is a slow release fertilizer because the organic constituents have to undergo several proteolytic changes by the soil bacteria before they become available to plants. Organic fertilizers, unlike their inorganic counterparts, release nitrogen only gradually and thus they will be available for plant growth perhaps over a season.

Fish offal and/or low value fish can be converted to fertilizer by digesting them with sulphuric acid. Acid digestion converts the proteins to ammonium sulphate and also makes the bone phosphate available for absorption by the plants.

As an alternate to acid digestion, another technique employed is to treat the fish with urea. The treatment solublises the fish tissues. One advantage of the process is that on manuring, the urea immediately becomes available for plant nutrition. Protein still in the organic form undergoes hydrolysis by the soil bacteria and becomes available for plant nutrition at a later time. Acid digestion as well as solubilisation with urea reduces the fish odour and make the handling of, and manuring with, the product not objectionable.

One of the shortcomings of a fertilizer processed according to the above method is the poor balance between components like nitrogen, phosphorus and potassium. This can be overcome if the fish material is digested with potassium hydroxide and the excess alkali is neutralised with phosphoric acid. This process will make up the deficiencies of potassium and phosphorus in the fish fertilizer.

19.19 CHITIN AND CHITOSAN

Chitin is the second most abundant organic compound on earth next only to cellulose. It is a white, hard, inelastic nitrogenous polysaccharide widely distributed in the exoskeletons of insects, crab, shrimp and lobster and also in the internal structures of other invertebrates, e.g. squid. Shrimp shell and head waste constitute the single largest source of chitin in India. Crab shell also is another major source. About 14-27 per cent of dry shrimp waste and 13-15 per cent of dry crab shell waste is chitin.

19.19.1 Chitin

19.19.1.1 Processing

The process for production of chitin from shrimp or crab waste involves two important steps viz., demineralisation of the waste using a mineral acid followed by deproteinisation of the demineralised mass using a dilute solution of caustic soda. Initial demineralisation reduces the bulk of the mass, particularly if dry shell waste is used, making the deproteinisation step simpler.

The shell waste is stirred well with dilute hydrochloric acid (1.2N) until it becomes soft. Demineralisation may take up to one hour for completion and is indicated by the stoppage of effervescence. The liquid is decanted and the mass is washed with water until free of acid. The demineralised mass is boiled in a 5 per cent solution of caustic soda in water for few minutes. The liquid portion containing the dissolved protein is filtered off and the residue is washed with water until free of alkali. The wet mass is dried under sun or in a hot air drier. The product is chitin. It is generally pulverised to the required mesh size before bagging.

19.19.2 Chitosan

Chitosan is deacetylated chitin. Whereas chitin is neutral and insoluble in water and most organic solvents, chitosan is basic in nature and is soluble in dilute acetic acid. However, chitosan processed from chitin from different sources and employing different processing parameters yield solutions in acetic acid showing varying viscosity. Therefore, selection of raw material and processing parameters employed are very significant. A general method is discussed below.

19.19.2.1 Processing

Chitin is mixed with a 40 per cent solution of caustic soda and heated by indirect steam at 95-100°C in a steam-jacketed kettle. Heating is continued for 90-120 minutes. During heating, samples are drawn at intervals from the reaction mixture, washed free of alkali and tested for its solubility in 1 per cent solution of acetic acid. Completion of deacetylation is indicated by complete solubility of the sample in acetic acid. At the end of the reaction, the caustic soda solution is drained off and the residue is washed repeatedly using water until it is free from alkali. The residue is then dried in sun or in a hot air drier. Chitosan is pulverised to the required mesh size before bagging.

19.19.3 Uses of Chitin and Chitosan

Chitin and chitosan are extensively used for several purposes. Chitin is a growth promoter in animals and birds and is used as an ingredient in their feed. Some of the important uses of chitosan are :

- as clarifying agent of fruit juices, and in purification of drinking water.
- as a thickening and stabilising agent in foods.
- in the treatment of waste water and sewage effluents.
- as sizing agent in textiles and paper.
- in cosmetics as a moisturiser and for protection against UV rays.
- as a base for chromatography.
- as a haemostatic agent in surgery and dentistry.
- in slow release of drugs.
- in fibres, films and membranes.

19.20 GLUCOSAMINE HYDROCHLORIDE

Another important derivative of chitin is glucosamine hydrochloride, which has applications in pharmaceuticals and as an additive in feeds.

19.20.1 Processing

One part by weight of pulverised chitin is mixed with three parts by volume of 10 N hydrochloric acid in a suitable reaction vessel fitted with a reflux condenser and a stirrer. It is heated in a boiling water bath for about two hours with stirring. When the chitin is completely dissolved in the acid, the excess acid is removed by a flash evaporator. Water is added to the slurry in the vessel and undissolved residue, if any, is filtered off. It is then mixed with little activated charcoal and set aside until the colour of the supernatant solution becomes pale yellow. Charcoal is filtered off and the clear liquid is concentrated using a flash evaporator. The crystals of glucosamine hydrochloride are collected and dried in air.

19.21 SQUALENE FROM SHARK LIVER OIL

Squalene is an unsaturated hydrocarbon present in the unsaponifiable fraction of fish oils. It is the main hydrocarbon found in the liver oil of certain species of shark and accounts for as high as 85 per cent of the unsaponifiable portion.

19.21.1 Isolation

Liver oil containing high proportion of squalene is taken in a stainless steel glass-lined vessel fitted with attachments for distillation under vacuum and is distilled under 2 mm bar pressure. Fraction distilling between 240 and 245°C is collected. All operations should preferably be carried out in an inert atmosphere because squalene is easily oxidised.

19.21.2 Uses

Squalene is used in the finishing operations of natural and artificial silks where it gives a brilliant sheen to the product. It is also used in perfumery as a carrier of perfumes. Other uses are in medicine, as lubricants and as bactericide.

19.22 AMBERGRIS

Ambergris is a fatty or pitch-like substance, gray or

black in colour, produced in the intestine of sperm whale, apparently when they are sick. Often, horny beaks of cuttlefish is found embedded in ambergris and therefore it is believed that ambergris forms around the irritating indigestible beaks of cuttlefish on which the sperm whales mainly feed. When the mass of ambergris become large and they are not got rid of the animal may die. It is the harder gray ambergris that accumulates this way causing the death of the animal. The softer, black variety is usually ejected by the whale as soon as it is formed. Fresh ambergris has strong and unpleasant smell. On exposure to air it hardens and develops a sweet musty odour. Ambergris is often found floating in tropical waters or cast on the seashore. It is found in lumps, very small to very large, weighing upto 100 kg.

19.22.1 Uses

Ambergris is widely used in the East as an aphrodisiac. Its main use is in perfumery as a fixative as it prevents the volatile oil from evaporating too quickly.

The foregoing provides a short account of the important byproducts processed out of fish and fish waste. However, this is not exhaustive, but is only indicative and scope is widening for processing newer products from fish and fishery waste.

Suggested Reading

Brody, J. (1965) *Fishery By-Products Technology*. The AVI Publishing Company, Westport, Connecticut, USA.

FAO (1974) *Production of Fishmeal and Oil*. FAO Fisheries Technical Paper No. 142. FAO, Rome.

Kreuzer, R. and Ahmed, R. (1978) *Shark Utilization and Marketing*. FAO International Trade Centre.

Pulikov, P.I. (1971) *Production of Meal, Oil and Protein-Vitamin Preparations in the Fishing Industry*. Amerind Publishing Co. Pvt. Ltd., New Delhi.

Stansby, M.E. (ed.) (1963) *Industrial Fishery Technology*. Robert E. Kreiger Publishing Company, New York.

Stansby, M.E. (1967) *Fish Oils*. The AVI Publishing Company, Westport, Connecticut, USA.

Subbarao, G.N. (1961) *Fisheries Products Manual*, FAO, Rome.

Suzuki, T. (1981) *Fish and Krill Protein Processing Technology*. Applied Science Publishers Ltd., London.

Wheaton, F.W. and Lawson, T.B. (1985) *Processing Aquatic Food Products*. Wiley Interscience Publication, New York.

Windsor, M and Barlow, S. (1981) *Introduction to Fishery By-Products*. Fishing News Books Ltd., Farnham, England.

Subject Index